# Catastrophe Theory

# V. I. Arnol'd

# Catastrophe Theory

### Third, Revised and Expanded Edition

Translated from the Russian
by G. S. Wassermann
Based on a Translation by R. K. Thomas

### With 87 Figures

Springer-Verlag
Berlin  Heidelberg  New York
London  Paris  Tokyo
Hong Kong  Barcelona
Budapest

Vladimir I. Arnol'd
Steklov Mathematical Institute
ul. Vavilova 42
Moscow 117966
Russia

Mathematics Subject Classification (1991): 00A06, 14J17, 30Fxx, 32Sxx, 34Axx, 34Cxx, 34Dxx, 35Bxx, 58Exx, 58Fxx, 58F14, 70Hxx, 70Kxx, 93Dxx

ISBN 3-540-54811-4 Springer-Verlag Berlin Heidelberg New York
ISBN 0-387-54811-4 Springer-Verlag New York Heidelberg Berlin

ISBN 3-540-16199-6 2. Auflage   Springer-Verlag Berlin Heidelberg New York
ISBN 0-387-16199-6 2nd Edition Springer-Verlag New York Heidelberg Berlin

Library of Congress Cataloging-in-Publication Data. Arnol'd, V. I. (Vladimir Igorevich), 1937– [Teoriia katastrof. English] Catastrophe theory / V.I. Arnol'd; translated from the Russian by G.S. Wassermann based on a translation by R.K. Thomas. -- 3rd rev. and expanded ed. p. cm. Translation of: Teoriia katastrof. Includes bibliographical references and index. ISBN 0-387-54811-4 (U.S.) 1. Catastrophes (Mathematics) I. Title.   QA614.58.A7613   1992   514' .74--dc20      92-9633

Production Editor: P. Treiber
Typesetting and printing: Zechnersche Buchdruckerei, Speyer
Bookbinding: J. Schäffer OHG, Grünstadt
41/3140-54321   Printed on acid-free paper

*To the Memory of M.A. Leontovich*

# Preface to the Third English Edition

The main difference between this edition and the previous English one is the inclusion of a short "textbook" on singularity theory at the end. It has the form of a sequence of problems or excercises, most of which are not too difficult, but a few of them are really hard. Readers may use them not only to check whether they have mastered the subjects discussed in the previous chapters, but also as a source of new information.

It is a pleasure to acknowledge the excellent work of the translator, G. Wassermann, who has corrected the original text trying to make it mathematically rigorous even when the author discusses skylarks, sycophants, atomic civil wars and other tragic mistakes of the past and the future.

March 1992                                      V. I. Arnol'd

# Preface to the Third Russian Edition

The mathematical description of the world depends on a delicate interplay between continuity and discontinuous, discrete phenomena. The latter are perceived first. "Functions, just like living beings, are characterised by their singularities", as P. Montel proclaimed. *Singularities, bifurcations* and *catastrophes* are different terms for describing the emergence of discrete structures from smooth, continuous ones.

In the last 30 years the theory of singularities has reached a highly sophisticated level, mainly due to the work of H. Whitney (1955), R. Thom (1959) and J. Mather (1965). It is a powerful new tool with a wide range of applications in science and engineering (especially when combined with bifurcation theory, which had its origins in the work of H. Poincaré ("Thèse", 1879) and was developed by A. Andronov (1933)).

The aim of this book is to explain how it works to a reader with no mathematical background. I hope, however, that even the experts will find here some facts and ideas new for them.

Some people consider catastrophe theory to be a part of the theory of singularities, while others, conversely, include singularity theory in catastrophe theory. To avoid any scholastical dispute, I will apply the term *catastrophe theorist* to anyone who himself claims to be working on catastrophe theory, thus leaving a free choice among "singularities", "bifurcations", or "catastrophes" to the authors of the publications discussed.

The first chapters of this book have appeared in the form of an article in the Moscow journal *Nature* (1979, issue 10), whose French translation with R. Thom's comments was published in Paris in 1980 (in *Matematica*). The Russian editions of 1981 and 1983 and the English editions of 1984 and 1986 each contained some new chapters. The present, most complete edition differs from the previous ones at many points. Information about the history of catastrophe theory has been added, and the chapters on geometric applications, on bifurcation theory and on applications to 'soft modelling', including the investigation of *perestroikas*, have been extended. Perhaps it is interesting to note that my attempts, beginning in 1986, to publish an analysis of perestroikas from the point of view of singularity theory have only now been crowned with success, undoubtedly as a consequence of *perestroika* itself.

Among the more mathematical questions included in the new edition, I shall note here the theory of delayed loss of stability, results on normal forms for implicit differential equations and relaxation oscillations, the theory of the interior scattering of waves in a nonhomogeneous medium, the theory of boundary singularities and imperfect bifurcations, the description of the caustic of the exceptional Lie group $F_4$ in terms of the geometry of a surface with boundary and the appearance of the symmetry group $H_4$ of the regular four-dimensional 600-hedron in variational calculus and optimal control problems, the theory of the metamorphoses of shock waves, the universality of cascades of doublings, triplings, etc.

The author thanks Prof. R. Thom, Prof. M. Berry and Prof. J. F. Nye for many useful suggestions and comments on the preceding editions of this booklet. According to Thom, the term "catastrophe theory" was invented by E. C. Zeeman, and the term "attractor" was first used by S. Smale (while the previous editions of this book give all the credit to Thom). Professor Nye has remarked that some very interesting topological reasons prevent the realisation of a number of metamorphoses of caustics (such as the birth of 'flying saucers') in optics, for caustics generated by the eikonal or Hamilton-Jacobi equation with a Hamiltonian which is convex with respect to momentum.

I learned singularity theory from a four-hour conversation with B. Morin after his inspired talk on Whitney and Morin singularities at the 1965 Thom seminar. Morin explained to me Mather's fundamental stability theorems, as announced by Mather in a then recent letter to Morin (and I found a different proof later the same day). Mather's unpublished 1968 paper on right equivalence was unfortunately (or fortunately) unknown to me, and I recognized the relationship of work similar to Mather's by Tjurina in 1967 (published in 1968) to my 1972 *"A, D, E"* paper dedicated to her memory only after J. Milnor explained it to me.

Neither in 1965 nor later was I ever able to understand a word of Thom's own talks on catastrophes. He once described them to me (in French?) as "bla-bla-bla", when I asked him, in the early seventies, whether he had proved his announcements. Even today I don't know whether Thom's statements on the topological classification of bifurcations in gradient dynamical systems depending on four parameters are true (in a corrected form, i.e. for generic metrics and potentials: a counterexample to the original "Thom theorem", as announced in *Topology* in 1969, was given by J. Guckenheimer in 1973, and even the "magic 7", so praised by catastrophe theorists, have to be augmented to make the theorem correct). The local topological classification of bifurcations in gradient dynamical systems depending on *three* parameters was recently obtained by B. A. Khesin (1985). The number of topologically different bifurcations proved to be finite, but significantly larger than had been conjectured by Thom, who had missed a number of bifurcations. Whether for *four* parameters the number of such bifurcations is *finite* (Thom claimed there were seven) is a question which has not yet been settled.

Nor am I able to discuss other, more philosophical or poetical declarations by Thom, formulated so as to make it impossible to decide whether they are true or false (in a manner typical rather of medieval science before Descartes or (the) Bacon(s)). Fortunately the fundamental mathematical discoveries of the great topologist do not suffer from any irrational philosophy.

As Poincaré once said, mathematicians do not destroy the obstacles with which their science is spiked, but simply push them toward its boundary. May these particular obstacles be pushed as far away as possible from this boundary, up to the domain of the unconscious and irrational.

Moscow, February 1986 and August 1991          V. I. Arnol'd

# Contents

# 1 Singularities, Bifurcations, and Catastrophes

The first information on catastrophe theory appeared in the western press in the late sixties. In magazines like "Newsweek" there were reports of a revolution in mathematics, comparable perhaps to Newton's invention of the differential and integral calculus. It was claimed that the new science, catastrophe theory, was much more valuable to mankind than mathematical analysis: while Newtonian theory only considers smooth, continuous processes, catastrophe theory provides a universal method for the study of all jump transitions, discontinuities, and sudden qualitative changes. There appeared hundreds of scientific and popular science publications in which catastrophe theory was applied to such diverse targets as, for instance, the study of heart beat, geometrical and physical optics, embryology, linguistics, experimental psychology, economics, hydrodynamics, geology, and the theory of elementary particles. Among the published works on catastrophe theory are studies of the stability of ships, models for the activity of the brain and mental disorders, for prison uprisings, for the behaviour of investors on the stock exchange, for the influence of alcohol on drivers and for censorship policy with respect to erotic literature.

In the early seventies catastrophe theory rapidly became a fashionable and widely publicized theory which by its all-embracing pretensions called to mind the pseudo-scientific theories of the past century.

The mathematical articles of the founder of catastrophe theory, René Thom, were reprinted as a pocket book – something that had not happened in mathematics since the introduction

1

of cybernetics, from which catastrophe theory derived many of its advertising techniques.

Shortly after the eulogies of catastrophe theory there appeared more sober critical works. Some of these also appeared in publications intended for a wide readership, under eloquent titles like – 'The Emperor has no clothes'. Now we already have many articles devoted to the criticism of catastrophe theory. (See for instance John Guckenheimer's survey article 'The Catastrophe Controversy' in 1978 and the parody on the criticism of the theory in 1979.)

The origins of catastrophe theory lie in Whitney's theory of singularities of smooth mappings and Poincaré and Andronov's theory of bifurcations of dynamical systems.

*Singularity theory* is a far-reaching generalization of the study of functions at maximum and minimum points. In Whitney's theory functions are replaced by mappings, i.e. collections of several functions of several variables.

The word *bifurcation* means *forking* and is used in a broad sense for designating all sorts of qualitative reorganizations or metamorphoses of various entities resulting from a change of the parameters on which they depend.

*Catastrophes* are abrupt changes arising as a sudden response of a system to a smooth change in external conditions. In order to understand what catastrophe theory is about one must first become acquainted with the elements of Whitney's singularity theory.

# 2 Whitney's Singularity Theory

In 1955 the American mathematician Hassler Whitney published the article 'Mappings of the plane into the plane', laying the foundations for a new mathematical theory – the theory of singularities of smooth mappings.

A *mapping* of a surface onto a plane associates to each point of the surface a point of the plane. If a point on the surface is given by coordinates $(x_1, x_2)$ on the surface, and a point on the plane by coordinates $(y_1, y_2)$ on the plane, then the mapping is given by a pair of functions $y_1 = f_1(x_1, x_2)$ and $y_2 = f_2(x_1, x_2)$. The mapping is said to be *smooth* if these functions are smooth (i.e. differentiable a sufficient number of times, such as polynomials for example.)

Mappings of smooth surfaces onto the plane are all around us. Indeed the majority of objects surrounding us are bounded by smooth surfaces. The visible contours of bodies are the projections of their bounding surfaces onto the retina of the eye. By examining the objects surrounding us, for instance, people's faces, we can study the singularities of visible contours.

Whitney observed that generically[1] only two kinds of singularities are encountered. All other singularities disintegrate under small movements of the body or of the direction of projection, while these two types are stable and persist after small deformations of the mapping.

An example of the first kind of singularity, which Whitney called a *fold,* is the singularity arising at equatorial points when a sphere is projected onto a plane (Fig. 1). In suitable

---

[1] I.e. for all cases bar some exceptional ones.

coordinates, this mapping is given by the formulas $y_1 = x_1^2$, $y_2 = x_2$. The projections of surfaces of smooth bodies onto the retina have just such a singularity at generic points, and there is nothing surprising in this. What is surprising is that besides this singularity, the fold, we encounter everywhere just one other singularity, but we practically never notice it.

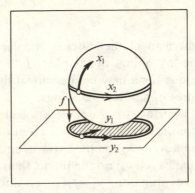

*Fig. 1.* The fold of the projection of a sphere onto the plane

This second singularity was named the *cusp* by Whitney, and it arises when a surface like that in Fig. 2 is projected onto a plane. This surface is given by the equation $y_1 = x_1^3 + x_1 x_2$ with respect to spatial coordinates $(x_1, x_2, y_1)$ and projects onto the horizontal plane $(x_2, y_1)$.

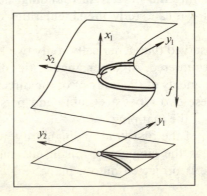

*Fig. 2.* The cusp of a projection of a surface onto the plane

4

Thus, the mapping is given in local coordinates by the formulas $y_1 = x_1^3 + x_1 x_2$, $y_2 = x_2$.

On the horizontal projection plane one sees a *semicubic parabola* with a cusp (spike) at the origin. This curve divides the horizontal plane into two parts: a smaller and a larger one. The points of the smaller part have three inverse images (three points of the surface project onto them), points of the larger part only one and points on the curve two. On approaching the curve from the smaller part, two of the inverse images (out of the three) merge together and disappear (here the singularity is a fold), and on approaching the cusp all three inverse images coalesce.

Whitney proved that the cusp is *stable*, i.e. every nearby mapping has a similar singularity at an appropriate nearby point (that is, a singularity such that the deformed mapping, in suitable coordinates in a neighbourhood of the point mentioned, is described by the same formulas as those describing the original mapping in a neighbourhood of the original point). Whitney also proved that *every singularity of a smooth mapping of a surface onto a plane, after an appropriate small perturbation, splits into folds and cusps.*

Thus the visible contours of generic smooth bodies have cusps at points where the projections have cusp singularities,

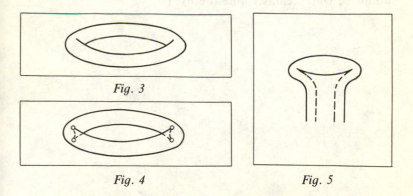

*Fig. 3*

*Fig. 4*

*Fig. 5*

*Fig. 3.* The visible contour of a torus
*Fig. 4.* The four cusps of the projection of a torus onto a plane
*Fig. 5.* Experimental verification of Whitney's theorem

and they have no other singularities. Looking closely, we can find these cusps in the lines of every face or object. Let us consider, for instance, the surface of a smooth torus (let us say, an inflated tyre). The torus is usually drawn as in Fig. 3. If the torus were transparent, we should see the visible contour depicted in Fig. 4. The corresponding mapping of the torus onto the plane has four cusp singularities. Thus, the ends of the lines of the visible contour in Fig. 3 are cusps. At these points the visible contour has a semicubic singularity.

A transparent torus is rarely seen. Let us consider a different transparent body – a bottle (preferably milk). In Fig. 5 two cusp points are visible. By moving the bottle a little we may satisfy ourselves that the cusp singularity is stable. So we have convincing experimental confirmation of Whitney's theorem.

After Whitney's basic work, singularity theory developed rapidly and is now one of the central areas of mathematics, where the most abstract parts of mathematics (differential and algebraic geometry and topology, the theory of reflection groups, commutative algebra, the theory of complex spaces, etc.) come together with the most applied (stability of the motions of dynamical systems, bifurcation of equilibrium states, geometrical and wave optics, etc.). E. C. Zeeman suggested that the combination of singularity theory and its applications should be called catastrophe theory.

# 3  Applications of Whitney's Theory

Since smooth mappings are found everywhere, their singularities must be everywhere also, and since Whitney's theory gives significant information on singularities of generic mappings, we can try to use this information to study large numbers of diverse phenomena and processes in all areas of science. This simple idea is the whole essence of catastrophe theory.

When the mapping we are concerned with is known in sufficient detail we have a more or less direct application of the mathematical singularity theory to various natural phenomena. Such applications do indeed lead to useful results, such as in the theory of elasticity and in geometrical optics (the theory of singularities of caustics and wave fronts, about which we shall say more later).

In the majority of works on catastrophe theory, however, a much more controversial situation is considered, where not only are the details of the mapping to be studied not known, but its very existence is highly problematical.

Applications of singularity theory in these situations are of a speculative nature; to give an idea of such an application we reproduce an example due (with slightly different details) to the English mathematician Christopher Zeeman of the speculative application of Whitney's theory to the study of the activity of a creative personality.

We shall characterize a creative personality (e.g. a scientist) by three parameters, called 'technical proficiency', 'enthusiasm', and 'achievement'. Clearly these parameters are interrelated. This gives rise to a surface in three-dimensional space with coordinates $(T, E, A)$.

Let us project this surface onto the $(T, E)$ plane along the $A$ axis. For a generic surface the singularities are folds and cusps (by Whitney's theorem). It is claimed that a cusp situated as indicated in Fig. 6 satisfactorily describes the observed phenomena.

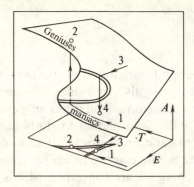

*Fig. 6.* The model of the scientist in 'technical proficiency-enthusiasm-achievement' space

In fact, let us see how under these assumptions the achievement of a scientist will change in dependence on his technical proficiency and enthusiasm. If enthusiasm is not great, then achievement grows monotonically and fairly slowly with technical proficiency. If enthusiasm is sufficiently great, then qualitatively different phenomena begin to occur. In this case with increasing technical proficiency achievement can increase by a jump (such a jump occurs for instance at point 2 in Fig. 6 as enthusiasm and technical proficiency change along curve 1). The domain of high achievement at which we then arrive is indicated in Fig. 6 by the word 'geniuses'.

On the other hand a growth of enthusiasm not supported by a corresponding growth in technical proficiency leads to a catastrophe (at the point 4 of curve 3 in Fig. 6) where achievement falls by a jump and we drop to the domain denoted in Fig. 6 by the word 'maniacs'. It is instructive that the jumps from the state of genius to that of maniac and back take place along different lines, so that for sufficiently great enthusiasm a genius and maniac can possess identical enthusiasm and tech-

8

nical proficiency, differing only in achievement (and previous history).

The deficiencies of this model and many similar speculations in catastrophe theory are too obvious to discuss in detail. I remark only that articles on catastrophe theory are distinguished by a sharp and catastrophic lowering of the level of demands of rigour and also of novelty of published results. If the former can be understood as a reaction against the traditional current in mathematics of rigorous but dull epigonic works, yet the careless treatment catastrophe theorists accord their predecessors (to whom the majority of concrete results are due) can hardly be justified. The motive in both cases is more social than scientific[1].

---

[1] "I think, my dearest, that all this decadence is nothing other than simply a way of approaching tradesmen." V. M. Doroshevich, *Rasskazy i Ocherki,* Moscow 1966, p. 295.

# 4 A Catastrophe Machine

In contrast to the example given above, the application of singularity theory to the study of bifurcations of equilibrium states in the theory of elasticity is irreproachably founded.

In many elastic constructions, under identical external loadings a number of different equilibrium states are possible. Consider, for instance, a horizontal rule whose ends are fixed by hinges and with a load at the centre.

Along with the equilibrium state where the rule sags under the load, a state is also possible in which the rule is curved upwards into an arch, not unlike a bridge.

With increase of load, at some point a 'catastrophe' or 'buckling' occurs: the rule jumps from one state to the other. Singularity theory can be applied to the study of such bucklings and its predictions stand up very well in experiments.

For visual illustration of this kind of application a number of devices have been invented: one of the simplest, called Zeeman's *catastrophe machine,* is illustrated in Fig. 7.

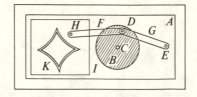

*Fig. 7.* Zeeman's catastrophe machine

The catastrophe machine can be easily made. One needs a board (A) (see Fig. 7) and a cardboard disk (B) secured to the board at its centre by a pin (C) so that it can rotate freely.

10

Another pin (D) is stuck only into the disk, at its edge, and a third (E) only into the board. To complete the assembly of the machine one needs two easily stretched rubber bands (F, G), a pencil (H) and a piece of paper (I).

After the pin on the edge of the disk has been linked to the fixed pin and the pencil by rubber bands, we place the point of the pencil somewhere on the sheet of paper and thereby stretch the rubber bands. The disk settles in a certain position. Now as we move the pencil the disk rotates. It turns out that in certain positions of the pencil a small change in its position can give rise to a 'catastrophe', i.e. the disk jumps to a new position. If we mark all these 'catastrophe positions' on the sheet of paper, we get the 'catastrophe curve' (K).

It turns out that the catastrophe curve obtained has four cusps. On crossing the catastrophe curve a jump may or may not take place, depending on the path taken by the pencil in going round the cusps.

By experimenting with this machine and trying to deduce the rule determining whether a jump takes place, the reader is easily convinced of the necessity of a mathematical theory of the phenomenon and can better appreciate the value of singularity theory in its explanation.

The states of the catastrophe machine are described by three numbers. The position of the point of the pencil is given by two coordinates (called the *control parameters*). The position of the disk is defined by one more number (the angle of rotation), also called the *internal parameter* of the system. If all three parameters are given, then the degrees of stretch of the rubber bands and consequently the potential energy of the whole system are determined. The disk rotates so as to minimize this energy (at least locally). For a fixed position of the pencil the potential energy is a function of the position of the disk, i.e. a function defined on the circle. This function can have one or several minima, depending on the values of the control parameters (Fig. 8, a). If on varying the control parameters the position of the minimum changes smoothly then there is no jump. A jump occurs for those values of the control parameters at which a local minimum disappears by combin-

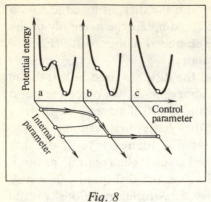
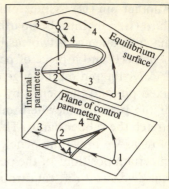

*Fig. 8*                                    *Fig. 9*

**Fig. 8.** The potential energy of the catastrophe machine
**Fig. 9.** The equilibrium surface of the catastrophe machine

ing with a local maximum (Fig. 8, b); after the jump the disk finds itself in a position corresponding to another local minimum (Fig. 8, c).

Let us examine the three-dimensional *state space* of the machine. The states at which the disk is in equilibrium form a smooth surface in this space. Let us project this surface onto the plane of the control parameters along the axis of the internal parameter (Fig. 9). This projection has folds and cusps. The projection of the fold points is just the catastrophe curve. Figure 9 shows clearly why a transition of the control parameters across the catastrophe curve sometimes causes a jump and sometimes does not. (It depends on which part of the surface the disk position corresponds to.) Using this diagram, one can get from one place on the equilibrium surface to another without jumps occurring.

The scheme of the majority of applications of catastrophe theory is the same as in the examples described. One assumes that the process under consideration can be described with the aid of a certain number of control and internal parameters. The equilibrium states of the process form a surface of some dimension in this space. The projection of the equilibrium surface onto the plane of control parameters can have singularities. One assumes that these are generic singularities. Then singularity theory predicts the geometry of the 'catastrophes', i.e. the jumps from one equilibrium state to another under a change of the control parameters. In the majority of serious applications, the singularity is a Whitney cusp and the result was known before the advent of catastrophe theory.

Applications of the type described are well-founded to the same degree as are their original premises. For instance, in the theory of buckling of elastic constructions and in the theory of capsizing of ships the predictions of the theory are completely confirmed by experiment. On the other hand, in biology, psychology, and the social sciences (for instance in applications to the theory of the behaviour of stock-market players or to the study of nervous illnesses), like the original premises, so the conclusions are more of heuristic significance only.

# 5 Bifurcations of Equilibrium States

An *evolutionary process* is described mathematically by a vector field in phase space. A point of phase space defines the *state* of the system. The vector at this point indicates the velocity of change of the state.

At certain points the vector may be zero. Such points are called *equilibrium positions* (the state does not change with time). Figure 10 shows the phase space of a system describing the interrelationship between predators and prey (say pike and carp). The phase space is the positive quadrant of the plane. The horizontal axis indicates the number of carp and the vertical axis the number of pike. $P$ is the equilibrium. The point $A$ corresponds to the equilibrium number of carp with a number of pike less than at equilibrium. Clearly in due course oscillations will set in; the equilibrium state in Fig. 10 is *unstable*. The oscillation is represented by a closed curve in the phase plane. This curve is called a *limit cycle*.

The curves in phase space traced by the successive states of a process are called *phase curves*. In a neighbourhood of a

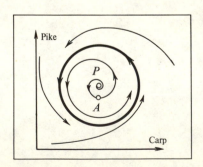

*Fig. 10.* The phase plane of the predator-prey model

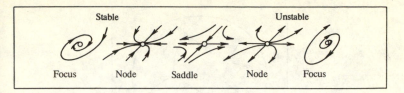

*Fig. 11.* The typical phase portraits in the neighbourhood of an equilibrium point

point that is not an equilibrium state, the partition of phase space into phase curves is just like a partition into parallel lines: the family of phase curves can be transformed into a family of parallel lines by a smooth change of coordinates. In a neighbourhood of an equilibrium point the picture is more complex. It was shown by Poincaré that the behaviour of phase curves in a neighbourhood of an equilibrium point on the phase plane of a generic system is as in Fig. 11. All more complicated cases turn into combinations of the ones shown after a small generic perturbation of the system.

Systems describing real evolutionary processes are as a rule generic. In fact, such systems always depend on parameters that are never known exactly. A small generic change of parameters transforms a system that is not generic into one that is.

So, generally speaking, all cases more complex than those indicated above should not be encountered in nature and at first glance can be neglected. This point of view invalidates a large part of the theory of differential equations and of mathematical analysis in general, where the main attention is traditionally paid to hard-to-investigate non-generic cases, of little real value.

However the situation is quite different if one is interested not in an individual system but in systems depending on one or more parameters. In fact, let us consider the *space of all systems* (Fig. 12), divided into domains of generic systems. The dividing surfaces correspond to degenerate systems; under a small change of the parameters a degenerate system becomes nondegenerate. A one-parameter family of systems is indicated by a curve in Fig. 12. This curve can intersect trans-

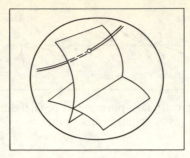

*Fig. 12.* A one-parameter family as a curve in the space of systems

versally (at a non-zero angle) the boundary separating different domains of nondegenerate systems.

Thus, although for each individual value of the parameter the system can be transformed by a small perturbation into a nondegenerate one, this cannot be done simultaneously for all values of the parameter: every curve close to the one considered intersects the dividing boundary at a close value of the parameter (the degeneracy, eliminated by the small perturbation at the given value of the parameter, appears anew at some nearby value).

And so *the degenerate cases are not removable when a whole family rather than an individual system is considered.* If the family is a one-parameter family then only the simplest degeneracies are unremovable, those represented by boundaries of codimension one (i.e. boundaries given by one equation) in the space of all systems. The more complex degenerate systems, forming a set of codimension two in the space of all systems, may be gotten rid of by a small perturbation of the one-parameter family.

If we are interested in two-parameter families then we need not consider degenerate systems forming a set of codimension 3, and so on.

This gives rise to a *hierarchy of degeneracies* by codimension and a *strategy* for their study: first we study generic cases, then degeneracies of codimension one, then two, etc. In this connection, *we must not restrict our study of degenerate systems to the picture at the moment of degeneracy, but must include a de-*

16

*scription of the reorganizations that take place as the parameter passes through the degenerate value.*

The above general considerations are due to H. Poincaré and are applicable not only to the study of equilibrium states of evolutionary systems but to a large part of mathematical analysis in general. Although they were proposed a hundred years ago, the progress in the realization of Poincaré's programme for a theory of bifurcation remains quite modest in most areas of analysis, partly due to the great mathematical difficulties and partly due to psychological inertia and the dominance of the axiomatic-algebraic style.

Let us return, however, to equilibrium states of evolutionary systems. At the present time we can regard as solved only the problem of the reorganizations of phase curves under bifurcations of equilibrium states in *one-parameter* generic families; even the case of *two* parameters is beyond the possibilities of present knowledge.

The results of the investigation of generic one-parameter families are summarized in Figs. 13–18. Figure 13 depicts a one-parameter family of evolutionary processes with a one-dimensional phase space (the parameter $\varepsilon$ is indicated along the horizontal axis and the state $x$ of the system along the vertical axis).

For a generic one-parameter family, the equilibrium states for all values of the parameter form a smooth curve ($\Gamma$ in Fig. 13; in the more general case, the dimension of the manifold of

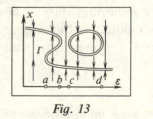

*Fig. 13*

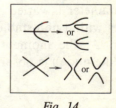

*Fig. 14*

*Fig. 13.* The curve of equilibria of a one-parameter family of systems

*Fig. 14.* The transformation of nontypical bifurcations into typical ones upon a small perturbation of the family

17

equilibrium states is equal to the number of parameters). In particular this means that the bifurcations depicted on the left-hand side of Fig. 14 will not occur in a generic family: a small perturbation of the family will transform $\Gamma$ into a smooth curve of one of the types[1] on the right-hand side of Fig. 14.

The projection of $\Gamma$ onto the parameter axis for a one-parameter family only has singularities of the fold type (for more parameters the more complex singularities of Whitney's theory appear: for instance, in generic two-parameter families the projection of the surface of equilibria $\Gamma$ onto the parameter plane can have cusp points, where three equilibrium states come together).

Thus, as we change the parameter we may single out singular or bifurcation values of the parameter (the critical values $a$, $b$, $c$, $d$ of the projection in Fig. 13). Away from these values the equilibrium states depend smoothly on the parameters. When the parameter approaches a bifurcation value an equilibrium state 'dies', by combining with another one (or, going the other way, a pair of equilibria is born 'out of thin air').

Of the two simultaneously appearing (or dying) equilibrium states one is stable and the other unstable.

At the moment of birth (or death) both the equilibrium states move with infinite speed: when the parameter value differs from the bifurcation value by $\varepsilon$ the distance between the two nearby equilibrium states is of the order of $\sqrt{\varepsilon}$.

Figure 15 depicts the metamorphosis of the family of phase curves on the plane in a generic one-parameter family. When the parameter is varied a stable equilibrium state (a 'node') collides with a non-stable one (a 'saddle') and then both disappear. At the moment of fusion a non-generic situation (a 'saddle-node') is observed.

---

[1] *Note:* 'type' here means equivalence class up to diffeomorphisms of the two-dimensional parametrized phase space as a whole, and not (which would be stronger) up to fibred diffeomorphisms (diffeomorphisms which respect the parametrization, i.e. which are themselves a parametrized family of diffeomorphisms of the one-dimensional phase space, together with a diffeomorphic change of parameter).

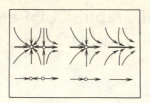

*Fig. 15.* The saddle-node: a typical local bifurcation in a
one-parameter family

In Fig. 15 it is clear that the metamorphosis is, in essence,
one-dimensional. Along the horizontal axis the same phenomena occur as on the $x$ axis in Fig. 13, but along the vertical axis
there is no reorganization at all. Thus, the metamorphosis
through a saddle-node is obtained from a one-dimensional
metamorphosis by 'suspending' it along the ordinate axis. It
turns out that in general all reorganizations of equilibrium
states in generic one-parameter systems can be obtained from
one-dimensional metamorphoses by means of an analogous
suspension.

If a stable equilibrium state describes the established conditions in some real system (say in economics, ecology or chemistry) then when it merges with an unstable equilibrium state,
the system must jump to a completely different state: as the
parameter is changed the equilibrium condition in the neighbourhood considered suddenly disappears. It was jumps of
this kind which lead to the term 'catastrophe theory'.

# 6 Loss of Stability of Equilibrium and of Self-Oscillating Modes of Behaviour

Loss of stability of an equilibrium state on change of parameter is not necessarily associated with the bifurcation of this equilibrium state: it can lose stability not only by colliding with another state, but also by itself.

The corresponding reorganization of the phase portrait on the plane is indicated in Fig. 16. Two versions are possible:

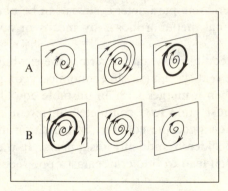

*Fig. 16.* Cycle-birth bifurcation

A. On change of the parameter the equilibrium state *gives birth to a limit cycle* (of radius of order $\sqrt{\varepsilon}$, where the parameter differs from the bifurcation value by $\varepsilon$). The stability of the equilibrium is transferred to the cycle, and the equilibrium point itself becomes unstable.

B. An *unstable limit cycle collapses* at the equilibrium state: the domain of attraction of the equilibrium state shrinks to nought with the cycle, which then disappears, transferring its instability to the equilibrium state.

It was observed by Poincaré and proved by Andronov and his pupils before the war (in 1939), that apart from the loss of stability of stable equilibrium states which merge with unstable ones (as described in Chap. 5), and the A and B cases just described, for generic one-parameter families of systems with a two-dimensional phase space no other forms of loss of stability are encountered. Later it was proved that also in systems having phase spaces of higher dimension loss of stability of equilibrium states on change of a single parameter must take one of the above forms (in the directions of all the additional coordinate axes the equilibrium remains attracting when the parameter is changed).

If our equilibrium state is the established behaviour in a real system, then under change of parameter the following phenomena are observed in cases A and B.

A. After loss of stability of the equilibrium *a periodic oscillatory behaviour is found to have become established* (Fig. 17), the amplitude of the oscillation being proportional to the square root of the criticality (the difference of the parameter from the critical value at which the equilibrium loses stability).

This form of loss of stability is called *mild* loss of stability since the oscillating behaviour for small criticality differs little from the equilibrium state.

B. Before the established state loses stability the domain of attraction of the state becomes very small and ever-present random perturbations throw the system out of this domain

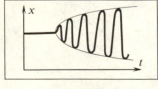

*Fig. 17*

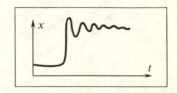

*Fig. 18*

*Fig. 17.* Mild loss of stability of an equilibrium
*Fig. 18.* Hard loss of stability of an equilibrium

21

even before the domain of attraction has completely disappeared.

This form of loss of stability is called *hard* loss of stability. Here the *system leaves its stationary state with a jump to a different state of motion* (see Fig. 18). This state can be another stable stationary state or stable oscillations, or some more complex motion).

The conditions of motion which establish themselves have recently come to be called attractors since they 'attract' neighbouring conditions (transients). [An *attractor* is an attracting set in phase space. Attractors that are not equilibrium states or strictly periodic oscillations have been given the name *strange attractors* and are connected with the problem of turbulence.]

The existence of attractors with exponentially diverging phase curves on them and the stability of such a kind of phenomenon was established at the beginning of the sixties in papers by S. Smale, D. V. Anosov, and Ya. G. Sinaj on the structural stability of dynamical systems.

Independently of these theoretical works, the meteorologist E. Lorentz in 1963 described an attractor in a three-dimen-

*Fig. 19.* A chaotic attractor

22

sional phase space which he had observed in computer experiments on modelling convection, with phase curves running apart on it in various directions (Fig. 19), and he pointed out the connection of this phenomenon to turbulence.

In Anosov's and Sinaj's papers exponential divergence was established in particular for the motion of a material point on a surface of negative curvature (a saddle is an example of such a surface). The first applications of the theory of exponential divergence to the study of hydrodynamic stability were published in 1966.

The motion of a fluid can be described as the motion of a material point on a curved infinite-dimensional surface. The curvature of this surface is negative in many directions, which leads to the rapid divergence of the trajectories, i.e. to poor predictability of the flow from the initial conditions. In particular, this implies the practical impossibility of a long-range dynamic weather forecast: for a prediction 1–2 months ahead one must know the initial conditions to an accuracy of $10^{-5}$ times the accuracy of the forecast.

However, let us return to the mode of behaviour which establishes itself following loss of stability of an equilibrium state and let us assume that this behaviour is a strange attractor (i.e. is not an equilibrium or a limit cycle).

The transition of a system to such a behaviour means that in it, complicated non-periodic oscillations are observed, the details of which are very sensitive to small changes of the initial conditions, while at the same time the average characteristics of the behaviour are stable and do not depend on the initial conditions (when they vary within some domain). An experimenter observing the motion of such a system would call it turbulent. It appears that the disordered motion of a fluid observed on loss of stability of laminar flow with an increase of the Reynolds number (i.e. with a decrease in viscosity) is described mathematically by just such complex attractors in the phase space of the fluid. The dimension of this attractor, it appears, is finite for any Reynolds number (for two-dimensional fluid flows Yu. S. Il'yashenko, M. I. Vishik and A. V. Babin recently obtained an upper bound for this dimension

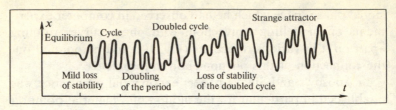

Fig. 20. A chaotisation scenario

with a magnitude of the order of $Re^4$), but tends to infinity as Re tends to infinity.

The transition from a stable equilibrium state of a process ('laminar flow of a fluid') to a strange attractor ('turbulence') can take place both by means of a jump (for hard or catastrophic loss of stability), or after a mild loss of stability (Fig. 20). In the latter case the stable cycle which was created itself then loses its stability. The loss of stability of a cycle in a generic one-parameter family of systems can take place in a number of ways: 1) *collision* with an unstable cycle (Fig. 21), 2) *doubling* (Fig. 22) and 3) the birth or death of a torus (Fig. 23) (in Andronov's terminology: *the skin peels off the cycle*). The details of these last processes depend on the resonances between the frequencies of the motion along the meridian of the torus and along its axis, i.e. on whether the ratio of these frequencies is rational or irrational. It is interesting that rational numbers with denominators 5 or higher behave practically like irrationals.

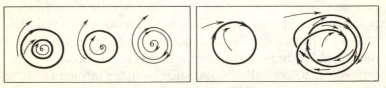

Fig. 21                                           Fig. 22

Fig. 21. The death of a cyclic attractor
Fig. 22. The doubling of an attracting cycle

24

*Fig. 23*                              *Fig. 24*

*Fig. 23.* Bifurcation involving the birth of a torus near a cycle
*Fig. 24.* Codimension 2 bifurcation near a 1:3 resonance

The behaviour of the phase curves close to a cycle can be described approximately with the aid of an evolutionary process for which the cycle is an equilibrium state. The approximative systems arising in this way are at present analyzed for all cases except those close to strong resonance with a frequency ratio of 1:4 (R. I. Bogdanov and E. I. Khorozov). Figure 24 shows the metamorphoses of the family of phase curves in an approximative plane system which correspond to the metamorphoses of the distribution of phase curves in the neighbourhood of a cycle in 3-space: it is assumed that the loss of stability occurs near a resonance of 1:3. Figure 25 shows one of the possible sequences of events near a resonance of 1:4. The basic results on this resonance were obtained not by strict mathematical arguments but by a combination of guesses and computer experiments (F. S. Berezovskaya and A. I. Khibnik, and A. I. Nejshtadt).

*Fig. 25.* One version of codimension 2 bifurcation near a 1:4 resonance

The Poincaré-Andronov theory set forth above on the loss of stability of equilibrium states has so many applications in all branches of oscillation theory (to systems with a finite number of degrees of freedom and also to continuous media) that it is impossible to enumerate them here: systems in mechanics, physics, chemistry, biology and economics lose stability all the time.

In articles on catastrophe theory mild loss of stability of equilibrium states is usually called Hopf bifurcation (partly through my 'fault' since, in talking of the Poincaré-Andronov theory to Thom in 1965, I particularly emphasised the work of E. Hopf, who had carried part of this theory over to the multidimensional case).

In bifurcation theory as in singularity theory the fundamental results and applications were obtained without the help of catastrophe theory. The undoubted contribution of catastrophe theory was the introduction of the term attractor and the spreading of knowledge on the bifurcation of attractors. A variety of attractors have been discovered now in all areas of oscillation theory; for instance it has been suggested that the different phonemes of speech are different attractors of a sound-producing dynamical system.

Under a slow change of parameter a qualitatively new phenomenon can be observed, the *delayed loss of stability* (Fig. 26).

After the parameter has passed through a bifurcational value corresponding to the birth of a cycle, i.e., corresponding to the mild arising of self-oscillations, the system remains in the neighbourhood of the equilibrium state which has lost stability for a certain while yet, during which the parameter has time to change by a finite amount. Only then does the system go over, by a jump, to the self-oscillating mode of behaviour born at the moment of bifurcation, so that the loss of stability appears to be hard.

It is interesting that this effect – the singularity of the dynamic bifurcation – takes place only in *analytic* systems. In the infinitely differentiable case the size of the delay of loss of stability, generally speaking, tends to nought as the speed of change of the parameter is decreased.

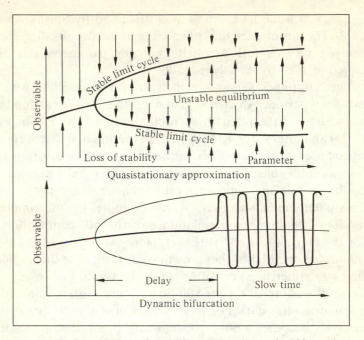

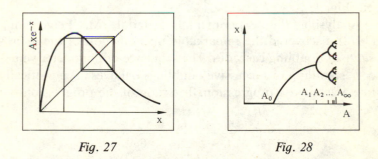

*Fig. 26.* Delay of loss of stability under dynamic bifurcation

*Fig. 27*                              *Fig. 28*

*Fig. 27.* Oscillations of population numbers in the simplest
Malthusian model taking competition into account

*Fig. 28.* A cascade of period doublings

Delay in a model example was described by Shishkova in 1973. The proof that this phenomenon occurs in all typical analytic systems with a slowly changing parameter was obtained in 1985 by A. I. Nejshtadt.

The humpback salmon catch is known to oscillate with a two-year period. An investigation of the ecological models called upon to explain these oscillations led A. P. Shapiro (1974) and afterwards R. May to the experimental discovery of *cascades of period doublings:* successive doubling bifurcations follow rapidly one after the other, so that an *infinite* number of doublings fall on a *finite* interval of change of the parameter. This phenomenon is observed, for example, for the simplest model of Malthusian population growth with competition – for the map $x \mapsto A x e^{-x}$ (Fig. 27). Here the factor $e^{-x}$, which diminishes the Malthusian growth coefficient $A$ as the population size $x$ increases, accounts for competition. For small values of the parameter $A$ the stationary point $x = 0$ is stable (the population dies out). For large values of $A$ the attractor successively becomes a nonzero stationary point (the bifurcation $A_0$), a cycle of period 2, Fig. 27, as for the humpback salmon (a doubling bifurcation, $A_1$), then one of period 4 ($A_2$), and so on (Fig. 28).

Analysing this experimental material, M. Feigenbaum (1978) discovered the remarkable phenomenon of the *universality* of doubling cascades. The sequence of parameter values corresponding to successive doublings behaves asymptotically like a geometric progression. The ratio of the progression

$$\lim_{n \to \infty} \frac{A_{n+1} - A_n}{A_n - A_{n-1}} \approx \frac{1}{4.669\ldots}$$

is a universal constant (not depending on the concrete system), like the numbers $\pi$ or $e$. The same cascades of doublings of limit cycles are also observed in typical evolutionary systems described by differential equations depending on a parameter.

Unlike period doubling, tripling is a phenomenon of codimension two. Cascades of triplings (and other period magnifications) become typical not in one-parameter, but in two-parameter families of systems. In these cases the universal indices are complex.

In the theory of two-parameter bifurcations considerable progress has been achieved in recent years. In particular, by 1987 H. Żołądek had solved the long-standing problem of the number of limit cycles born out of the zero equilibrium position in systems of the Volterra-Lotka type (Fig. 10) described by vector fields on the plane tangent to the sides of an angle.

However, the bifurcation problem in the system

$$\dot{z} = \varepsilon z + A z^2 \bar{z} + \bar{z}^3,$$

to which the investigation of loss of stability of self-oscillations reduces in the only remaining unexplored case of codimension 2, still has not yielded to the efforts of mathematicians. In the plane of the complex parameter $A$ one can single out 48 regions (Fig. 29) in which bifurcations occur differently when the small complex parameter $\varepsilon$ moves around zero. (It has not even been proved that the full number of such regions is finite, although it is conjectured that there are 48 in all.)

Even recently every experimenter finding complicated aperiodic oscillations, say in a chemical reaction, rejected

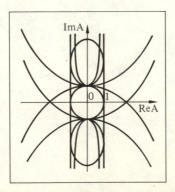

*Fig. 29.* Forty-eight types of codimension 2 bifurcation at a 1:4 resonance

them from consideration, citing impurities in the experiment, chance external effects and such like. Now it is clear to many that these complex oscillations may be connected with the very essence of the matter, may be determined by the fundamental equations of the problem and not by random external effects; they can and must be studied on a level with the classical stationary and periodic modes of behaviour of processes.

# 7 Singularities of the Stability Boundary and the Principle of the Fragility of Good Things

Let us consider an equilibrium state of a system depending on several parameters and let us assume that (in some domain of variation of the parameters) this equilibrium state does not bifurcate.

We shall represent the system corresponding to some value of the parameters by a point on the parameter axis (on the plane if there are two parameters, in the parameter space if there are three, and so on).

The domain being studied in the parameter space then splits into two parts corresponding to whether the equilibrium state is stable or not. Thus we obtain on the plane (in the space) of parameters a *stability domain* (consisting of those values of the parameters for which the equilibrium is stable), an *instability domain,* and separating them the *stability boundary.*

In accordance with Poincaré's general strategy (see Chap. 5) we shall restrict ourselves to families of systems depending on parameters in a generic way. It turns out that the stability boundary can have singularities that do not disappear with small perturbations of the family.

Figure 30 depicts all singularities of the stability boundary of equilibrium states in generic two-parameter families of evolutionary systems (with phase spaces of arbitrary dimension), Fig. 31 for three-parameter families. The formulas in the diagrams describe the stability domain (under a suitable choice of coordinates in the parameter plane or space, generally, curvilinear ones).

We remark that *in all cases the stability domain is arranged 'with corners outward', so driving wedges into the instability domain.* Thus for systems belonging to the singular part of the

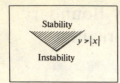

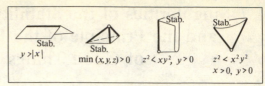

*Fig. 30*                                    *Fig. 31*

*Fig. 30.* The typical singularity of the boundary of a
two-dimensional stability domain

*Fig. 31.* The typical singularities of the boundaries of
three-dimensional stability domains

stability boundary a small change of the parameters is more likely to send the system into the unstable region than into the stable region. This is a manifestation of a general principle stating that all good things (e.g. stability) are more fragile than bad things.

It seems that in good situations a number of requirements must hold *simultaneously,* while to call a situation bad *even one* failure suffices.

In the four-parameter case we must add two more to the boundary singularities enumerated above. As the number of parameters increases, the number of types of singularities of the stability boundaries of generic families rapidly grows. However, it was shown by L. V. Levantovskij that the number of singularity types (not reducible to each other by smooth changes of parameters) remains finite for an arbitrarily large number of parameters, and the principle of fragility is also retained.

# 8 Caustics, Wave Fronts, and Their Metamorphoses

One of the most important deductions of singularity theory is the *universality* of certain simple forms like folds and cusps which one can expect to encounter everywhere and which it is useful to learn to recognise. As well as the previously enumerated singularities, one often meets some further types, which have been given their own names: the *'swallowtail'*, the *'pyramid'* (which Thom calls the *'elliptic umbilic'*), the *'purse'* (which Thom calls the *'hyperbolic umbilic'*), and so on.

Suppose that a disturbance (e.g. a shock wave, light, or an epidemic) is being propagated in some medium.

For simplicity let us start with the plane case. Let us suppose that at the initial moment of time the disturbance is on the curve *a* (Fig. 32) and that the speed of its propagation is 1. To find out where the disturbance will be at time *t* we must lay out a segment of length *t* along every normal to the curve. The resulting curve is called the *wave front*.

Even if the initial wave front has no singularities, after some time singularities will begin to appear. For instance, upon

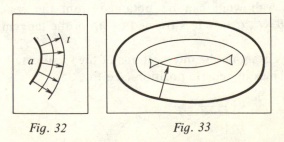

Fig. 32          Fig. 33

*Fig. 32.* The evolution of a wave front
*Fig. 33.* The singularities of the equidistants of an ellipse

33

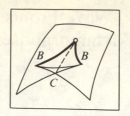

*Fig. 34.* The swallowtail

propagation of a disturbance inside an ellipse the singularities indicated in Fig. 33 appear. These singularities are stable (not removable by small perturbations of the initial wave front). For a generic smooth initial front, as time goes on only standard singularities of this type will arise.

All other singularities (e.g. the singularity at the centre of a shrinking circle) decompose on a small perturbation of the initial front into several singularities of standard type.

On a generic smooth wave front in three-dimensional space, with the passage of time only cusp ridges and standard singularities of the *swallowtail* type shown in Fig. 34 can appear (try to picture the singularities of a wave front being propagated inside a triaxial ellipsoid). All more complex singularities dissolve on small perturbations of the front into swallowtails joined by cusp ridges and self-intersection curves.

The swallowtail can be defined as the set of all points $(a, b, c)$ such that the polynomial $x^4 + ax^2 + bx + c$ has a multiple root. This surface has a cusp ridge ($B$ in Fig. 34) and a curve of self-intersection ($C$ in Fig. 34).

The swallowtail can be obtained from the space curve $A = t^2$, $B = t^3$, $C = t^4$: it is formed by all of the tangents to this curve.

Let us examine the intersections of the swallowtail with parallel planes in generic position (see Fig. 35).

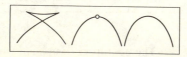

*Fig. 35.* The typical metamorphosis of a wave front on the plane

These intersections are plane curves. As the plane is translated these curves change form at the moment the plane passes through the vertex of the tail. The transformation (*metamorphosis*) here is exactly the same as the metamorphosis of a wave front on a plane (for instance during the propagation of a disturbance inside an ellipse).

We can describe the metamorphoses of wave fronts in the plane as follows. Side by side with the basic space (the plane in this case) let us also consider *space-time* (three-dimensional in the present case). The wave front being propagated in the plane sweeps out a surface in space-time. It turns out that this surface can itself always be regarded as a wave front in space-time ('*the big front*'). In the generic case, the singularities of the big front will be swallowtails, cusp ridges and self-intersections, situated in space-time in a generic way relative to the isochrones (which are made up of 'simultaneous' points in space-time). Now it is easy to understand which metamorphoses can be experienced by the momentary wave fronts on the plane in the generic case: they are the changes in form of the isochronic cross-sections of the big front.

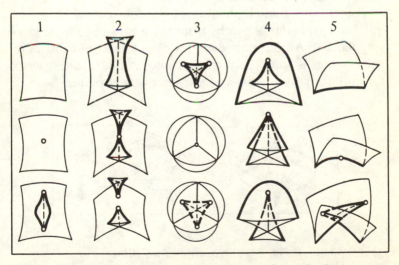

*Fig. 36.* The typical metamorphoses of wave fronts in three-space

The study of the metamorphoses of a wave front during its propagation in three-dimensional space leads in the same way to an investigation of the cross-sections of the big (three-dimensional) wave front in four-dimensional space-time by the three-dimensional isochrones. The metamorphoses which arise are illustrated in Fig. 36.

The study of the metamorphoses of wave fronts was one of the problems out of which catastrophe theory arose: however, even in the case of three-dimensional space, catastrophe theorists could not manage this problem; Fig. 36 appeared only in 1974 when new methods (based on the theory of crystallographic symmetry groups) were developed in singularity theory.

Along with wave fronts, *ray systems* can be used to describe the propagation of disturbances. For example the propagation of a disturbance inside an ellipse can be described using the family of internal normals to the ellipse (Fig. 37). This family has an envelope. The envelope of the family of rays is called a *caustic* (i.e. 'burning', since light is concentrated at it). A caustic is clearly visible on the inner surface of a cup when the sun shines on it. A rainbow in the sky is also due to a caustic of a system of rays that have passed through a drop of water with complete internal reflection (Fig. 38).

The caustic of an elliptical front has four cusps. These singularities are *stable*: a nearby front has a caustic with the same singularities. All the more complex singularities of caustics resolve under a small perturbation into standard ones: cusps (given locally by $x^2 = y^3$) and self-intersection points.

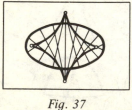

*Fig. 37* 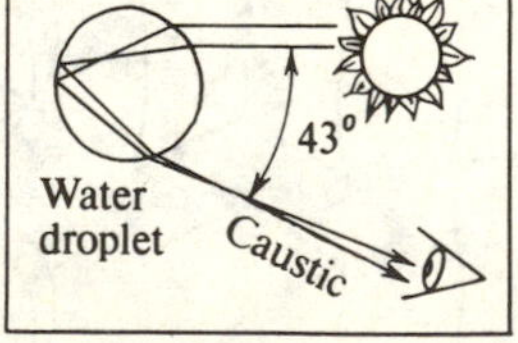                    *Fig. 38*

*Fig. 37*. The caustic of an ellipse
*Fig. 38*. Descartes' theory of the rainbow

The system of normals to a surface in three-dimensional space also has a caustic. This caustic can be constructed by marking off on each normal the radius of curvature (a surface, in general, has two different radii of curvature at each point so that the normal has two caustic points).

It is not easy to imagine the caustics of even the simplest surfaces, a triaxial ellipsoid for instance.

Generic caustics in three-dimensional space have only standard singularities. These singularities are called the *'swallowtail'*, the *'pyramid'* or *'elliptic umbilic'*, and the *'purse'* or *'hyperbolic umbilic'* (see Fig. 39). The pyramid has three cusp ridges meeting tangentially at the vertex. The purse has one cusp ridge and consists of two symmetric boat bows intersecting in two lines. These singularities are stable.

*All more complex singularities of caustics in three-dimensional space resolve into these standard elements on small perturbations.*

Let us consider, for one initial front (for instance an ellipse in the plane), both its caustic and the fronts of the disturbance being propagated. It is not difficult to see that *the singularities of the propagating front slide along the caustic and fill it out*.

For instance the metamorphosis of wave front 5 in Fig. 36 corresponds to a swallowtail on the caustic. The cusp ridge of the wave front moving in three-dimensional space sweeps out the surface of the caustic (a swallowtail). However, this partitioning of the caustic into curves is *not the same* partitioning of the swallowtail surface into plane curves that we encountered earlier (Fig. 35). The cusp ridge of the moving front does not have self-intersections. The cusp ridge of the moving front

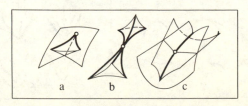

*Fig. 39.* The typical singularities of caustics in three-space

37

passes twice through each point of the self-intersection line of the caustic. The time interval between these passings is very small (of the order of $\varepsilon^{5/2}$, where $\varepsilon$ is the distance from the vertex of the tail).

In exactly the same way, in the metamorphoses 3 and 4 of Fig. 36 the cusp ridges of the moving fronts sweep out the pyramid and the purse.

If the original front is moved (under control of a parameter) then its caustic moves also and during this movement can undergo metamorphoses. *The metamorphoses of a moving caustic on the plane can be studied by considering cross-sections of a big caustic in space-time,* as was done for fronts. The metamorphoses obtained are depicted in Fig. 40. (These are the metamorphoses of plane sections of the swallowtail, purse and pyramid). All more complex metamorphoses decompose into sequences of these ones on a small perturbation of a one-parameter family.

Let us turn our attention to the metamorphosis 1 representing the formation of a caustic 'out of thin air'. The newly-formed caustic has the form of a sickle (*'lips'* in Thom's terminology) with semicubic cusps at the ends. In a similar way the visible contour of a surface can appear 'out of thin air' when the direction of projection changes (Fig. 41). Looking at a mound from above we do not see a contour. When the line of vision tilts, first a point singularity appears which then rapidly grows (proportionally to $\sqrt{t - t_0}$, where $t_0$ is the time at which

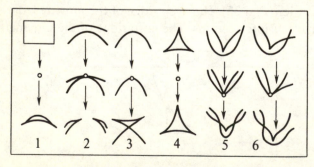

*Fig. 40.* The typical metamorphoses of caustics on the plane

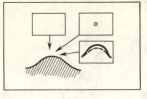  

Fig. 41                    Fig. 42

*Fig. 41.* The 'lip' metamorphosis: the birth of a visible contour
*Fig. 42.* The metamorphosis of a plane section of a cusp-ridged surface

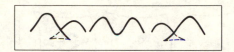

*Fig. 43.* The 'camel' metamorphosis

the singularity appears) and takes the form of 'lips'. The metamorphosis described here can be realized as the metamorphosis of a plane section of a cusp-ridged surface under translation of the plane (at the moment the form changes the plane is tangent to the cusp ridge (Fig. 42)).

Metamorphosis 3 can also be seen on a visible contour – one need only look at a two-humped camel while walking past it (Fig. 43). At the moment of metamorphosis the profile has the same singularity as the curve $y^3 = x^4$.

All metamorphoses of visible contours of surfaces in generic one-parameter families are covered by the first three illustrations in Fig. 40, 1–3.

*The metamorphoses of caustics moving in three-dimensional space* are obtained by means of the cross-sections of the big (three-dimensional) caustics in four-dimensional space-time by the three-dimensional isochrones. These metamorphoses are shown in Figs. 44 and 45.

One of these metamorphoses (1) describes the formation of a new caustic 'out of thin air'. We see that the newly-formed caustic has the appearance of a saucer with a sharp rim. At

39

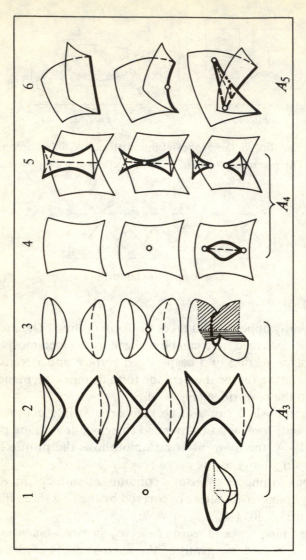

*Fig. 44.* The typical metamorphoses of caustics in three-space: the
*A* series

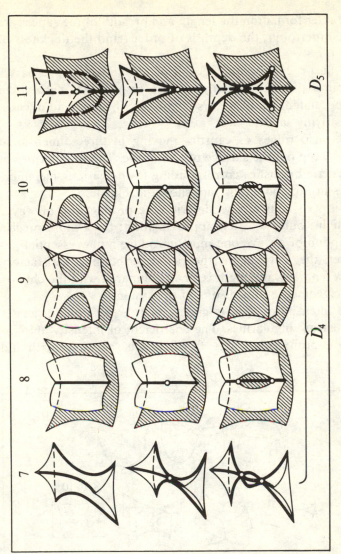

*Fig. 45.* The typical metamorphoses of caustics in three-space: the *D* series

time $t$ after formation the length and breadth of the saucer are of the order of $\sqrt{t}$, the depth is of order $t$ and the thickness of order $t\sqrt{t}$.

A caustic can become visible when a light beam passes through a dispersive medium (dust, fog). V. M. Zakalyukin has speculated that observers might describe this kind of caustics as flying saucers.

The cusp ridges of caustics moving in three-dimensional space sweep out the surface of a *bicaustic*. The singularities of the generic bicaustics corresponding to the various metamorphoses in Figs. 44 and 45 are shown in Fig. 46.

As is well known, rays describe the propagation of waves (say, light) only as a first approximation; for a more precise description of a wave one must introduce a new essential parameter, the wavelength (the ray description is satisfactory only when the wavelength is small compared with the characteristic geometric dimensions of the system).

The intensity of light is greater near a caustic and greater still near its singularities. The coefficient of intensity amplification is proportional to $\lambda^{-\alpha}$, where $\lambda$ is the wavelength and

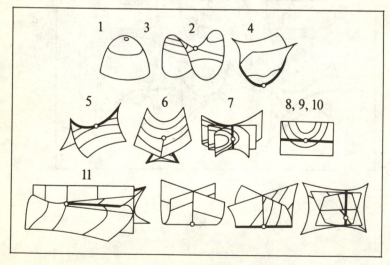

*Fig. 46.* The typical singularities of bicaustics

42

the index $\alpha$ is a rational number depending on the nature of the singularity. For the simplest singularities the values of $\alpha$ are as follows:

| caustic | cusp ridge | swallowtail | pyramid | purse |
|---------|------------|-------------|---------|-------|
| 1/6 | 1/4 | 3/10 | 1/3 | 1/3 |

Thus the brightest-shining are the point singularities of pyramid and purse type. In the case of a moving caustic, at isolated moments in time the even brighter singularities $A_5$ and $D_5$[1] can appear (see Figs. 44, 45; $\alpha = 1/3$ for $A_5$ and $3/8$ for $D_5$).

If the light is so intense that it can destroy the medium, then the destruction will start at the points of greatest brightness, and so the index $\alpha$ determines how the intensity for destruction to occur depends on the frequency of the light.

A classification of singularities of caustics and wave fronts analogous to the one described above has been carried out in multidimensional spaces up to dimension 10 (V. M. Zakalyukin).

The predictions by the theory of singularities of the geometry of caustics, wave fronts and their metamorphoses have been completely confirmed in experiments and it seems strange now that this theory was not constructed two centuries ago. However, the fact is that the mathematical apparatus needed is not trivial[2] and is connected with such diverse areas of mathematics as the classification of simple Lie algebras and of Coxeter's crystallographic groups, the theory of braids, the theory of the ramification of integrals depending on parameters, etc. It is even linked (in a quite mysterious way) to the classification of regular polyhedra in three-dimensional Euclidean space.

---

[1] All the singularities listed above belong to the two families $A_k$ and $D_k$, which are discussed in greater detail in Chap. 16.

[2] The original proof of Whitney's theorem, with which we began, was about forty pages long; while one can easily understand and use the final geometrical results of singularity theory, the proofs still remain complicated.

Catastrophe theorists try to avoid serious mathematics. For instance Zeeman's extensive 1980 bibliography on catastrophe theory omits references to most of the mathematical papers published after 1976. As a result, catastrophe theorists go on trying to discover experimentally answers to problems already solved by mathematicians. For instance, in a 1980 paper on wind fields and the movement of ice one finds semi-successful attempts to guess the list of the metamorphoses of caustics in three-dimensional space (see Figs. 44 and 45), published by mathematicians back in 1976.

# 9 The Large-Scale Distribution of Matter in the Universe

At the present time the distribution of matter in the universe is highly inhomogeneous (there are planets, the sun, stars, galaxies, clusters of galaxies and so on). Astrophysicists nowadays think that in the early stages of the development of the universe there was no such inhomogeneity. How did it come about? Ya. B. Zel'dovich in 1970 proposed an explanation of the formation of clusters of dustlike material that is mathematically equivalent to the analysis of the formation of singularities of caustics begun in 1963 by E. M. Lifshitz, Khalatnikov and Sudakov.

Let us consider a collision-free medium, i.e. a medium so rarified that its particles pass 'through' one another without colliding. For simplicity let us assume that the particles do not interact and move under inertia: in the time $t$ a particle starting at a point $x$ goes to the point $x + vt$.

Let us assume that at the initial moment of time the velocity of a particle at the point $x$ is $v_0(x)$; the vector field $v_0$ is called the *initial velocity field* of the medium. As time goes by the particles move and the velocity field changes (though the velocity of each particle does not change, at each succeeding moment of time this particle will be at a new position).

In Fig. 47 we show an initial velocity field $v_0$ of a homogeneous medium and the fields $v_1$, $v_2$, $v_3$ obtained from it after the times $t = 1, 2, 3$. We see that from a certain moment on *the faster particles begin to leave the slower ones behind,* with the result that the velocity field becomes three-valued: through one point of space three particle streams pass with different velocities.

45

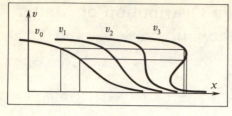
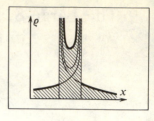

*Fig. 47*                                   *Fig. 48*

*Fig. 47*. The evolution of the velocity field of a collision-free
medium

*Fig. 48*. The singularities of the density after an overtaking

The movement of our medium can be described as a *one-parameter family of mappings of the line onto the line*. That is, for each $t$ there is a mapping $g_t$ taking the initial position of each particle ($x$) to the final position $g_t(x) = x + v_0(x) t$.

The mapping $g_0$ is the identity mapping (the transformation that takes each point to itself). For $t$ close to 0 the mappings are one-to-one and have no singularities. After the moment of the first overtaking, $g_t$ has two folds.

Suppose that at the beginning the density of the medium at a point $x$ is $\rho_0(x)$. The density changes with time. It is not difficult to see that after an overtaking the density graph will have the form shown in Fig. 48 (at a distance $\varepsilon$ from the fold point, the density is of the order $1/\sqrt{\varepsilon}$).

Thus, small deviations from constancy in the initial velocity field lead after long enough times to the formation of accumulations of particles (at places of infinite density).

This conclusion still holds when one goes from a one-dimensional medium to a medium filling a space of any dimension, and when one allows for the effects on the motion of particles of an external force field or a field originating from the medium, and also when the effects of relativity and the expansion of the universe are accounted for.

If the force field has a potential (i.e. the work done in moving along any path depends only on the beginning and end of the path), and the initial velocity field is also a potential field, then the problem of describing the singularities of the mappings $g_t$ and their metamorphoses under a change of $t$ is mathematically identical to that of describing the singularities of caustics and their metamorphoses (both are the subject of the theory of the so-called *Lagrangian singularities*.)

In the case of a two-dimensional medium, the points of infinite density form curves on the plane. These curves are formed by the critical values of the mapping $g_t$, i.e. the values at the critical points (for the mapping in Fig. 1 the critical points are the points on the equator of the sphere, the critical values are the points of the visible contour on the horizontal plane).

The curve of critical values of the mapping $g_t$ is called its *caustic*. It can be defined as the set of points where two infinitely close rays (particle trajectories) intersect, i.e. in the same way as the usual optical caustic.

In the same manner, the description of the metamorphoses of optical caustics carries over to the metamorphoses of particle clusters (points of infinite density of the medium) in potential motion.

The first singularity on the plane looks like a sickle with semicubically pointed tips (in three-dimensional space a newly-born caustic has the form of a saucer). Ya. B. Zel'dovich called such caustics *pancakes* ('bliny' in Russian; at first pancakes were interpreted as galaxies, later as clusters of galaxies).

As the medium continues its motion, new pancakes arise. Also, the existing pancakes begin to transform and may interact with one another. A typical sequence of events in a two-dimensional medium is illustrated in Fig. 49.

All the elementary metamorphoses that are possible in a three-dimensional medium are illustrated in Figs. 44 and 45

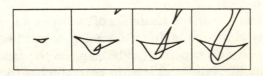

*Fig. 49.* The scenario of the interaction of Zel'dovich's 'pancakes'

(these results require the complicated mathematical theory of Lagrangian singularities).

As a result of the metamorphoses the density has singularities of different orders on the surface of the pancakes, along the curves, and at the isolated points. We shall characterize a singularity by the *mean density* in an $\varepsilon$-neighbourhood of the point under consideration (i.e. the ratio of the mass in an $\varepsilon$-neighbourhood to the volume of the neighbourhood.)

At points on the caustic the mean density tends to infinity as the radius $\varepsilon$ of the neighbourhood tends to zero.

The order of magnitude of the mean density at various points of the caustic is as follows:

| caustic | cusp ridge | swallowtail | purse, pyramid |
|---|---|---|---|
| $\varepsilon^{-1/2}$ | $\varepsilon^{-2/3}$ | $\varepsilon^{-3/4}$ | $\varepsilon^{-1}$ |

With changing time the singularities $A_5$ (mean density of order $\varepsilon^{-4/5}$) and $D_5$ ($\varepsilon^{-1} \ln 1/\varepsilon$) can appear at isolated moments.

According to astrophysicists, when the radius of the universe was a thousand times smaller than at present the large-scale distribution of matter in the universe was practically uniform and the velocity field was practically potential. The subsequent movement of particles led to the formation of caustics, i.e. singularities of the density and clusters of particles. Up to the formation of pancakes the density remains so small that the medium can be considered to be collision-free. After this moment, the medium can be considered collision-free if one assumes that mass-bearing neutrinos account for the significant part of the mass of the universe; if, however, most of the mass resides in protons and neutrons then deductions from the geometry of caustics and their metamorphoses must be treated with caution since the medium then ceases to be collision-free.

The deductions on the clustering of matter on surfaces, with the primary clustering taking place along certain lines (cords) meeting at particular points (nodes), appear to agree with astronomical observations at least in general terms (S. F. Shandarin).

# 10 Singularities in Optimization Problems: the Maximum Function

Many singularities, bifurcations, and catastrophes (jumps) arise in all problems in which extrema (maxima and minima) are sought, problems in optimization, control theory and decision theory. For instance, suppose we have to find $x$ such that the value of a function $f(x)$ is maximal (Fig. 50). Under a smooth change of the function the optimal solution changes with a jump from one of the two competing maxima ($A$) to the other ($B$).

Below we shall consider a number of problems of this type, all far from being completely solved, although in some cases adequate classifications of the singularities are known.

Let us consider a *family* $f(x, y)$ of functions of the variable $x$, parametrized by $y$. For each fixed value of $y$ let us compute the maximum of the function, denoting it by

$$F(y) = \max_x f(x, y).$$

The function $F$ is continuous but not necessarily smooth, even when $f$ is a polynomial.

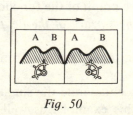
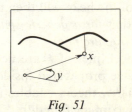

*Fig. 50*          *Fig. 51*

*Fig. 50.* The discontinuity of the optimal direction

*Fig. 51.* A fracture of the horizon line of a smooth landscape

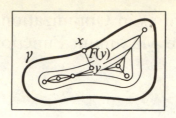

*Fig. 52.* The distance to a curve and its singular points

*Example 1.* Let $y$ be the azimuth of the line of vision, $x$ be the distance and $f$ be the angular elevation of the landscape at distance $x$ in the azimuth direction $y$ (Fig. 51). Then $F$ determines the *line of the horizon*. It is clear that *the horizon of a smooth surface can have fractures and that these cannot be removed by small perturbations*.

The variable $x$ and the parameter $y$ can be points in spaces of any dimension; along with maxima, minima are also encountered.

*Example 2.* Let $x$ be a point on a plane curve $\gamma$, let $y$ be a point in the region bounded by this curve and let $f(x, y)$ be the distance from $y$ to $x$.

We shall consider $f$ as a function of the point on the curve, depending on the point of the region as a parameter. Then the minimum function of the family, $F(y)$, is the shortest distance from the point $y$ to the curve $\gamma$ (Fig. 52). It is clear that this function is continuous but not everywhere smooth.

We may imagine a *spade*, bounded by the curve $\gamma$; let us fill the spade with as large a heap of dry sand as is possible. The surface of the heap will then be the graph of the function $F$. It is clear that *for a generic spade the surface of the heap has a ridge (a fracture line)*.

The level lines of $F$ are nothing other than the fronts of a disturbance propagated inside $\gamma$.

Singularity theory enables us to enumerate the singularities of the maximum function $F$ in this example, as well as for generic families of functions of any number of variables as long as the number of parameters $y$ is not greater than 10

(L. N. Bryzgalova). Let us consider the simplest cases of one and two parameters.

Choosing suitable coordinates on the $y$-parameter axis (or plane), and after subtracting from $F$ a suitable smooth function of the parameter(s), we may reduce the maximum function of a generic family in a neighbourhood of each point to one of the following normal forms:

one parameter: $F(y) = |y|$;

two parameters: $F(y) = \begin{cases} |y_1| & \text{or} \\ \max(y_1, y_2, y_1 + y_2) & \text{or} \\ \max_x(-x^4 + y_1 x^2 + y_2 x). \end{cases}$

The formula for the one-parameter case shows, in particular, that the horizon of a generic smooth landscape has no singularities other than the simplest fractures. The singularities of the maximum function described by the formulas for the two-parameter case give the following singularities for a minimum function (for instance, the singularities of a heap of sand on a spade): a ridge line, a point of conjunction of three ridges, and the end of a ridge (see Fig. 52).

In the last case the graph of the minimum function is the part of the swallowtail surface (see Fig. 34) obtained by removing the pyramid ($BCB$) adjoining the cusp ridge (and reflecting the surface in Fig. 34 in a horizontal plane).

For 3, 4, 5, and 6 parameters the number of different singularities is 5, 8, 12, and 17 respectively. From 7 parameters upwards, the number of non-equivalent singularity types becomes infinite: the normal forms inevitably contain 'moduli', which are functions of additional continuous parameters.

Topologically, the maximum (minimum) function of a generic family of smooth functions has the local structure of a generic smooth function on the parameter space (V. I. Matov).

In Fig. 53 we depict the typical singularities of the *locus of nonsmoothness of the maximum function* of a three-parameter family.

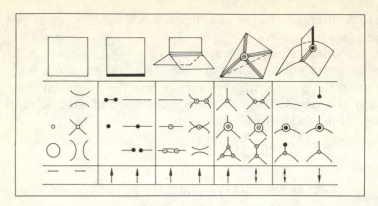

Fig. 53. The typical singularities of the nonsmoothness locus of
the maximum and the typical metamorphoses of shock waves

They allow us to investigate the typical metamorphoses of
singularities of *shock waves* on the plane taking place with the
passage of time: for this it is necessary first to study the typical
metamorphoses of two-dimensional sections of the five surfaces shown in Fig. 53 (these metamorphoses are also dis-
played in the drawing). It turns out that some of them do and
some of them do not represent metamorphoses of shock waves
(for example, for the potential solutions of the Burgers equation $u_t + u u_x = \varepsilon u_{xx}$ with vanishing viscosity $\varepsilon$).

Namely, the metamorphoses which can be realised by shock
waves are those marked by arrows in Fig. 53. The selection
principles were found by I. A. Bogaevskij and Yu. M. Barysh-
nikov:

1) the shock wave arising after the metamorphosis is contractible in a neighbourhood of the metamorphosis point;

2) at the moment of metamorphosis and immediately after it
the complement of the shock wave is topologically (homotopi-
cally) the same.

Each of these conditions is necessary and sufficient for a
typical metamorphosis of singularities of a maximum function
to be realisable by a typical metamorphosis of shock waves on
the plane and in three-space. Whether this is so in the multidimensional case is not known.

# 11 Singularities of the Boundary of Attainability

A *controlled system* in phase space is defined as follows: at every point of the space we have not just one velocity vector (as in the usual evolutionary system), but a whole set of vectors called the *indicatrix of permissible velocities* (Fig. 54).

The control problem is to choose at each moment of time a velocity vector from the indicatrix so as to reach a given target (for example to reach some subset of phase space in the shortest possible time).

The dependence of the shortest time to reach the target upon the initial point can have singularities. The singularities considered in Chap. 10 of the minimum function for the distance to a curve are a particular case (there the indicatrix is a circle and the target is the curve). In contrast to this particular case, the singularities of the shortest time in the general control problem have been studied very little.

In the general case it may not be possible to reach the target under all initial conditions. The points of the phase space

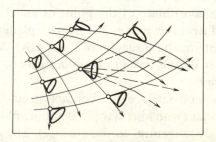

*Fig. 54.* The field of indicatrices of permissible velocities of a controlled system

53

from which the target can be reached (in any time) are called *the domain of attainability*.

The boundary of the attainability domain can have singularities even when neither the target nor the field of the indicatrices at the various points of phase space have them. We shall give below a classification of the singularities of the attainability boundary in a generic controlled system on a phase plane for the case where the indicatrices and the target are smooth curves (due to A. A. Davydov).

Of the four types of boundary singularities, three are given by simple formulas (under an appropriate choice of local coordinates on the plane):

$$1)\ y=|x|$$
$$2)\ y=x|x|$$
$$3)\ y=x^2|x|$$

The singularity of the fourth type is connected with the theory of differential equations not resolved with respect to the derivative, which are also called *implicit differential equations*.

Such an equation has the form $F(x,y,p)=0$, where $p=dy/dx$. Geometrically the equation $F=0$ defines a surface in the three-dimensional space with coordinates $(x,y,p)$. It is called the *equation surface*.

The condition $p=dy/dx$ singles out a plane at each point of our three-dimensional space. This plane consists of the vectors whose $y$ component is $p$ times bigger than the $x$ component, where $p$ is the coordinate of the point of attachment. Such a plane is called a *contact plane*. The contact plane at each point is vertical (contains the direction of the $p$ axis). All the contact planes together define a *field of contact planes,* also called a *contact structure*.

On the equation surface the contact structure cuts out a field of directions (with singular points at the places where the contact plane is tangential to the surface). The equation surface is here assumed to be smooth. This condition is fulfilled for generic equations.

The question of the structure of the typical singular points of implicit differential equations was already considered in the last century, and King Oscar II of Sweden included it, together with the three-body problem, in the list of four problems for the prize of 1885.

The solution of this problem was obtained only in 1985 by A. A. Davydov as a by-product of an investigation of the domains of attainability of controlled systems on the plane.

The answer is furnished by the following list of normal forms (to which the equation can be reduced by a local diffeomorphism of the plane):

$$y = (x + kp)^2.$$

Depending on the value of the parameter $k$, three cases are possible here. The singular point of the field on the equation surface can turn out to be a saddle, a focus or a node. The projection mapping of the equation surface onto the $(x, y)$-plane along the $p$ axis has a fold as a singularity. In a neighbourhood of a typical point of the fold the equation reduces to the normal form of M. Cibrario (1932), $x = p^2$. All singular points automatically hit the fold. The result of folding is depicted in Fig. 55: the singular points on the $(x, y)$-plane are called the *folded saddle* (*focus, node* respectively). It turns out that despite the complexity of the pattern formed by the inte-

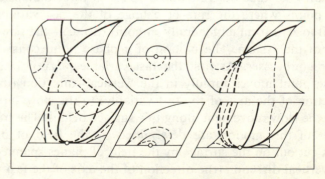

*Fig. 55.* Folded singularities

gral curves on the $(x, y)$-plane, it is uniquely determined (not only topologically, but even up to diffeomorphisms) by a single 'modulus' $k$ (as is also the phase portrait of the corresponding vector field on the plane near the singular point).

Folded singular points – saddles, foci, nodes – are encountered in many applications. Let us consider, for example, the *asymptotic lines* on a surface in three-space (the surface has higher than first-order tangency with the tangent line at each point of these curves). For a generic surface the network of asymptotic lines fills out the domain of hyperbolicity, where the surface has negative curvature (like an ordinary saddle). Through each point of the domain of hyperbolicity two asymptotic lines pass. The domain of hyperbolicity is bounded by the curve of parabolic points, upon which both asymptotic directions coincide.

In the neighbourhood of a typical parabolic point the asymptotic lines have a semicubic singularity and the whole network of them reduces to the same normal form $y = c \pm x^{3/2}$ as does the family of integral curves of the Cibrario equation.

However, in the neighbourhood of isolated points on the curve of parabolicity the behaviour of the asymptotic lines is more complicated: they are arranged like the integral curves of implicit equations near folded singular points (Fig. 55).

Folded singularities also appear in the theory of relaxation oscillations. Suppose the system has one fast and two slow variables, so that the full phase space is three-dimensional. The points where the rate of change of the fast variable is equal to zero form a (generally smooth) surface – the *slow surface* of the system. The motion of a phase point consists of several processes. At first the fast variable *relaxes,* i.e., the phase point moves rapidly in the 'vertical' direction (in the direction of the axis of the fast variable) to the slow surface, then the slow movement along this surface begins. The trajectories of this motion are determined by the field of directions cut out on the surface by the field of planes spanned by the vertical direction (the direction of the axis of the fast variable) and the direction of the perturbations. Generally speak-

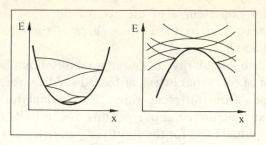

*Fig. 56.* Energy dissipation in a well and at a barrier

ing this field of planes defines a contact structure in the phase space, and the singularities of the slow motion are described, in general, by the folded singular points of Fig. 55.

As the last example of these singularities let us consider the motion of a mass point in a potential well (or near a potential barrier) in the presence of friction. Let us denote (Fig. 56) by $x$ the coordinate of the point and by $E$ its total energy. The projections of the phase trajectories onto the $(x, E)$-plane have in general semicubic singularities upon approaching the graph of the potential energy. To a minimum (maximum) of the potential energy there corresponds a singular point. For a generic potential energy one obtains all of the same Davydov folded singularities.

The reason why the folded singularities are encountered so often is that one often encounters both ordinary singularities of a vector field and foldings. What is unexpected here is merely that the combination of folding with singularity does not lead to a greater variety of cases than in the classification of vector field singularities themselves. Namely, let us consider folding as an involution (a diffeomorphism whose square is the identity transformation) on a plane carrying a vector field with a singular point. If the curve of fixed points of the involution passes through the singular point of the field and on this curve the involution changes the sign of each vector of the field to its opposite, then such an involution can (almost always) be converted to any neighbouring involution with such properties by the aid of a diffeomorphism which moves each phase curve of the given field along itself. This (rather unexpected) result is the source of the entire theory described above.

The fourth singularity of the attainability boundary is obtained out of two separatrices of folded saddles which enter a folded node from different sides. The normal form of the folded node mentioned above permits one to write down explicitly a normal form for the fourth singularity, but I shall not do this here.

Examples of controlled systems and targets that have attainability boundary singularities of these types are illustrated in Figs. 57, 58 and 59. In these diagrams the target $\gamma$ is denoted by a double line, the attainability boundary by a $T$-dashed line (the stem of the $T$ points into the attainability domain). The curves with the arrows are tangent to the edges of the cones of permissible directions at each point: the horizontally hatched area is the domain of 'complete controllability' (the convex hull of the indicatrix encloses 0). By looking at Figs. 57–59, the reader can convince himself that the four singularities are not removable.

To aid in understanding these drawings, we shall construct a network of *limiting curves,* defined as follows:

At each point outside the domain of complete controllability the directions of the permissible velocities lie inside an angle of less than 180°.

The sides of this angle determine the directions of the *limiting velocities* at the given point. Thus at each point outside the domain of complete controllability there are two limiting directions. The integral curves of the fields of limiting directions (i.e. curves that have a limiting direction at each of their points) are called *limiting curves*.

The network of limiting curves is indicated in Fig. 54 together with the indicatrices of permissible velocities (they have the shape of ellipses) and the angles based on the indicatrices, formed by the permissible directions of motion.

The boundary of the attainability domain consists of segments of limiting curves (and possibly segments of the target curve, when the target does not lie within the region of com-

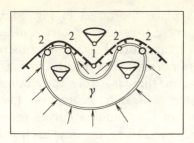

*Fig. 57.* The stability of the singularities 1 and 2 on the boundary
of the attainability domain

plete controllability, see Fig. 57). These segments meet each
other at certain points, and it is these which constitute the singularities of the attainability domain boundary.

In Fig. 57 the target has the form of the contour of a letter $C$
lying on its back. The permissible velocities are the same at all
points of the plane and they are directed upwards at an angle
comprising at most 45° with the vertical.

The inclination of all limiting curves is ±45°. The attainability boundary is denoted by the $T$-dashed line. It is evident
that the singular points on the boundary are of two types,
1 and 2.

At point 1 *segments of two different limiting curves meet*.
They intersect at a non-zero angle. It is clear that from points
lying above the boundary indicated in Fig. 57, it is not possible to reach the target by motion in a direction making an angle of 45° or less with the vertical, whereas for points under-
neath, it is possible. It is interesting to note that at vertex 1 the
domain of attainability yawns open: the unattainability domain drives a wedge into the domain of attainability at this
point. Thus the good region in the sense of Chap. 7 turns out
to be just the unattainable one.

At a point 2 on the attainability boundary *a segment of the
limiting curve meets a segment of the target curve*. At this point
the direction of the target curve is limiting, so that the attainability boundary has a tangent. However, at a point 2 the curvature of the boundary jumps as one passes from the limiting
curve to the target curve.

Now let us replace the target in Fig. 57 by an arbitrary nearby smooth curve (nearness of curves means the nearness of their tangents, curvatures, etc.) and let us replace the field of indicatrices of permissible velocities in Fig. 57 by a nearby field. Then it is clear that, as before, the attainability boundary of the new system will have, near point 1, a fracture point (where segments of two limiting curves meet at a non-zero angle). In exactly the same way, near each point 2 a point of analogous character will arise for the new system.

Thus the situation depicted in Fig. 57 is stable with respect to small perturbations of the system. The situations depicted in Figs. 58 and 59 have a similar stability property. The events which lead to the singularities of networks of limiting curves which are shown in these diagrams are as follows:

In Fig. 58 the curve $K$ bounds the hatched region of complete controllability: in the hatched area motion in any direction is possible (if so-called mixed strategies are allowed, i.e. motions which constantly change tack). The target in Fig. 58 lies within the domain of complete controllability. Consequently, the entire domain bounded by $K$ is attainable.

On the boundary $K$ of the region of complete controllability the permissible directions subtend an angle of exactly 180°. The boundary $K$ is formed by those points in the plane for which the double tangent which convexifies the indicatrix of permissible velocities passes through the origin of the velocity

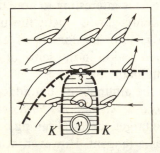

*Fig. 58.* The stability of the singularity 3 on the boundary of the attainability domain

60

plane (a double tangent is a straight line which is tangent to the curve at two points).

In Fig. 58 this double tangent is horizontal at every point of $K$. The circumstance leading to the formation of the singularity represented in Fig. 58 is that here *the curve $K$ is itself tangent to the double tangent to the indicatrix passing through 0.*

For a generic system such an event can only occur at isolated points of the boundary $K$ of the domain of complete controllability. In Fig. 58 it happens at point 3, where the tangent to $K$ is horizontal.

From what was said above it is clear that the circumstance described is realized in a stable way: under a small perturbation of the system, i.e. of the target and of the indicatrix field, point 3 will move slightly but will not vanish.

Let us now consider the network of limiting curves near the point 3. Both fields of limiting directions are smooth nearby. By a choice of a suitable system of coordinates we may straighten one of them. In Fig. 58 such a system has already been chosen: the first of the two families of limiting curves consists of horizontal lines (directed to the left).

The lines of the second family are smooth curves. Along $K$ they are tangential to the lines of the first family. At the point of interest 3 both families are tangent to the curve $K$. From these considerations it is not difficult to see now that the network of limiting curves near the point 3 looks just as indicated in Fig. 58: above the curve $K$ the lines of the second family rise as one moves in the allowed direction, whereas below $K$

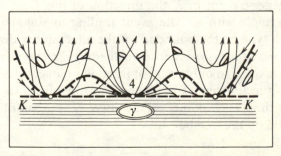

*Fig. 59.* Folded singularities on the boundary of the attainability domain

they fall (the choice of the directions of the lines of the network admits a few further variants analogous to the one depicted; the reader who has understood Fig. 58 can easily sort them out for himself).

It is now evident in Fig. 58 that to the left of point 3 the attainability boundary follows the line of the second limiting direction, and to the right it follows the line of the first (horizontal) one. At point 3 the two curves are tangent of the second order (being a straight line and a cubic parabola). In a neighbourhood of this point the attainability boundary is diffeomorphic[1] to the curve $y = x^2|x|$.

Thus, points 1 and 2 of Fig. 57 and point 3 of Fig. 58 give examples of stably realized circumstances on the boundary of the attainability domain which call forth singularities of the three simple types on p. 54. Singularities of the fourth type arise in the situation shown in Fig. 59.

In this drawing, just as in Fig. 58, the target is situated within the hatched domain of complete controllability. The points of the plane at which the convex indicatrix passes through zero lie on the boundary $K$ of this region. It is clear that in generic controlled systems this occurrence (the indicatrix passing through zero) is realized along a curve. On one side of this curve $K$ lies the domain of complete controllability (the indicatrix surrounds zero), and on the other, the region with two limiting directions. On the separating boundary $K$, these two fields of directions merge into one – the field of directions tangent to the indicatrices at zero.

At a generic point of $K$ the direction of this field makes a non-zero angle with $K$. The event leading to singularities of the fourth type on the boundary of the attainability domain is the *tangency of the curve $K$ with the limiting direction*. In a generic system this can only occur at isolated points of the boundary $K$ of the domain of complete controllability. In Fig. 59 there are three such points on the curve $K$; the middle one is designated by the number 4.

---

[1] A diffeomorphism is a change of coordinates which is smooth together with its inverse.

In order to study the network of limiting curves in the neighbourhood of these singular points, it is useful to consider our two-valued field of limiting directions as a single-valued field of directions on a surface which doubly covers the region above the curve $K$.

To this end let us consider the set of all directions of line elements on the plane. This set is a three-dimensional manifold, since a direction is determined by the point of attachment of the line element (two coordinates) and by its azimuth (one angular coordinate).

The set of all limiting directions forms a subset of the set of all directions. This subset is a smooth surface in the three-dimensional manifold of all directions. The three-dimensional manifold of all directions projects down onto the original plane (each line element projects onto its point of attachment). The surface formed by the limiting directions is mapped under this projection into the part of the plane lying above the curve $K$. This projection mapping of the surface onto the plane has a singularity over the curve $K$, namely a Whitney fold.

The two-valued field of limiting directions on the plane defines a single-valued field of directions everywhere on the constructed surface except at just those singular points of the curve $K$ (where the indicatrix is tangent at 0 to $K$) which we wish to study.

After being transported to the constructed surface, the limiting curves of both fields of limiting directions form a system of phase curves of a smooth vector field with singularities at the points which interest us. These singular points can be nodes, foci, or saddles (in Fig. 59 the middle point is a node and the two outer ones are saddles). Thus, the distribution of the limiting curves on the original plane can be obtained from the distribution of the phase curves of a vector field in the neighbourhood of a singular point by applying a Whitney fold mapping. Although this Whitney mapping and the phase curves are not completely independent (in particular, over $K$ the phase curves are tangential to the kernels of the projection), enough has been said to enable one to investigate the arrangement of limiting curves near a singular point (incidentally, the same picture is given by the asymptotic lines near a parabolic curve on a surface).

Figure 59 illustrates one of the variants of this situation. In the drawing it is evident that the $T$-dashed attainability boundary is formed by the projections of the separatrices of the saddles (the outer singular points) under the mapping of the two-fold covering surface onto the plane. On the covering above point 4 there lies a singular point of the 'node' type. Two separatrices of saddles come into this node from different directions.

At the node these two curves have a common tangent and (in the generic case) can be represented in a suitable coordinate system by the equations of parabolas of degree $\alpha > 1$, of the form

$$y = A|x|^{\alpha} \text{ for } x \leq 0, \qquad y = B|x|^{\alpha} \text{ for } x > 0 .$$

The fourth singularity of the attainability boundary is obtained from this pair of parabolas of degree $\alpha$ on the covering surface under a Whitney fold mapping.

Incidentally, this circumstance shows the error of the vulgar interpretation, extremely widespread among catastrophe theorists, of R. Thom's statement that "in nature one meets only stable phenomena and therefore in every problem one should study the stable cases, rejecting the others as not being realized". In the present case the singularities of the first three types are stable (up to equivalence by diffeomorphisms), but the fourth is not. Nevertheless all four types of singularities are encountered equally often and the study of the last is in no way less important than the investigation of the other three.

About the singularities of the domain of attainability, of the time function and the optimal strategy in generic controlled systems with a phase space of large dimension astonishingly little is known – only in 1982 was it proved (by A. A. Davydov) that the attainability domain is a topological manifold with a Lipschitz boundary.

One of the intermediate questions in the study of controlled systems is that of the singularities of convex hulls of smooth manifolds (curves, surfaces, and so on).

The *convex hull* of a set is the intersection of all half-spaces containing it. The indicatrix in a controlled system need not be convex.

However, it turns out that a non-convex indicatrix can be replaced by its convex hull.

For instance, the indicatrix of velocities of a yacht with the wind fore is not convex (Fig. 60). It is possible to move against the wind, however, by *changing tack*, and applying a *mixed strategy*, that is, alternating stretches of motion at different velocities belonging to the indicatrix. The mean velocity of motion under a mixed strategy belongs to the set of arithmetic means of the indicatrix vectors used, i.e. to the convex hull.

Singularities of convex hulls of generic curves and surfaces in three-dimensional space have been studied by V. D. Sedykh and V. M. Zakalyukin. For curves, up to a smooth change of variables the hull is given in a neighbourhood of each of its points by one of the six formulas:

$$z \geq 0, \qquad z \geq |x|, \qquad z \geq x|x|,$$
$$z \geq \min(u^4 + xu^2 + yu), \qquad z \geq \min^2(x, y, 0),$$
$$\{z \geq \min^2(x, y, 0), \ x + y \geq 0\}$$

(Fig. 61). For surfaces it is given by one of the three formulas:

$$z \geq 0, \ z \geq x|x|, \ z \geq \rho^2(x, y),$$

where $\rho(x, y)$ is the distance from the point $(x, y)$ to the angle $y \geq c|x|$ (Fig. 62). The number $c > 0$ is a modulus (invariant):

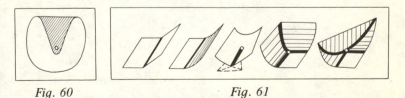

*Fig. 60*        *Fig. 61*

*Fig. 60.* Convexification of the indicatrix with the aid of a mixed strategy

*Fig. 61.* The typical singularities of convex hulls of space curves

65

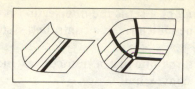

*Fig. 62.* The typical singularities of convex hulls of surfaces

hulls corresponding to different $c$'s cannot be transformed into one another by a smooth transformation.

Singularities of convex hulls in higher-dimensional spaces have been studied little. According to Sedykh the convex hull of a generic $k$-dimensional manifold in a space of dimension greater than $k+2$ has moduli that are functions of $k$ variables.

The shadow cast by an infinitely smooth or even an analytic convex body may, however strange it might seem, have singularities. Namely, the boundary of the shadow of a three-dimensional convex body can have discontinuities of the third derivative, and for a body of dimension 4 and above, even of the second derivative (I. A. Bogaevskij, 1990).

Many new and interesting singularities arise in optimization problems with constraints, for instance in the problem of by-passing obstacles. Their investigation led to new results in one of the most classical areas of mathematics – the geometry of smooth surfaces in three-dimensional space.

# 12 Smooth Surfaces and Their Projections

A smooth curve on the plane can have a tangent which touches it tangentially at any number of points (Fig. 63), but this is not true for generic curves. By a small perturbation of the curve one can always achieve that no straight line will be tangential at more than two points.

At how many points can a straight line be tangential to a generic surface? After some thought or experimentation, the reader can convince himself that *the maximum number of points of tangency is four*; one can move a line while maintaining three points of tangency, and while keeping two one can move it in two directions.

The *order of tangency* of a straight line with a curve or a surface can also vary (for example the order of tangency of the *x*-axis with the graph of $y = x^2$ is one, for $y = x^3$ two etc.). A generic plane curve does not have tangents of order greater than two (the second order of tangency is encountered at isolated points of the curve, called *points of inflection*).

For a surface in space the matter is not so simple. At points near which the surface is not convex (or concave), there are

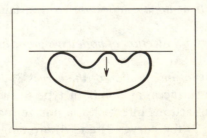

*Fig. 63.* A threefold tangent of a nontypical curve

67

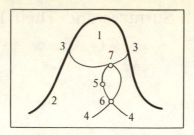

*Fig. 64.* The classification of points on a smooth surface

tangents of higher than first order (they are called *asymptotic tangents*). For generic surfaces there can be tangents of third order along certain curves, and of fourth order at isolated points; tangents to a generic surface of order greater than four do not exist.

All points of a generic surface can be divided into the following seven classes according to the orders of their tangents (Fig. 64):

1) The *domain of elliptic points* (all tangents are of order 1);

2) The *domain of hyperbolic points* (two asymptotic tangents).

These two domains are separated by a common boundary:

3) The *curve of parabolic points* (one asymptotic tangent).

Within the domain of hyperbolicity there is a special curve:

4) The *curve of inflection of the asymptotic lines* (where there is a tangent of third order).

Finally, on this curve are singled out some more special isolated points of three types:

5) The *points of double inflection* (with a tangent of fourth order);

6) The *points of inflection of both asymptotic lines* (two third-order tangents);

7) The *points of intersection of the curves* 3) *and* 4).

For generic surfaces, at points of type 6) two branches of the curve of inflections intersect at a non-zero angle, while at points of type 7) the curves 3) and 4) are tangential (of first order).

This classification of the points of a surface (due to O. A. Platonova and E. E. Landis) is connected in the following way with the classification of singularities of wave fronts.

Mathematicians speak of *points* to talk about objects of an arbitrary nature. Let us consider, for example, the set of all non-vertical straight lines on the $(x, y)$-plane.

Such lines are given by equations of the form $y = ax + b$. Consequently, one line is determined by a pair of numbers $(a, b)$ and can be regarded as a point of the plane with coordinates $(a, b)$. This plane is called the *dual to the original plane*. Its points are the straight lines of the original plane.

Given a smooth curve on the original plane, then at each of its points there is a tangent line. As a point moves along the curve, the tangent changes, and consequently, a point moves in the dual plane. Thus in the dual plane we get a curve – the set of all the tangents of the original curve. This curve is called the *dual curve to the original one*.

If the original curve is smooth and convex, then the dual curve is also smooth, and if the original curve has a point of inflection, then on the dual curve there is a corresponding *cusp* (Fig. 65).

Curves dual to generic smooth curves have the same singularities as generic wave fronts on the plane, and they undergo the same metamorphoses under a generic smooth deformation of the original curve as occur in the generic propagation of a generic wave front in the plane.

In exactly the same way the planes in three-dimensional space form a *dual three-dimensional space* and all the tangent planes to a smooth surface form a *dual surface*. The singularities of a surface dual to a generic surface are the same as those of a generic wave front, i.e. cusp ridges with swallowtails.

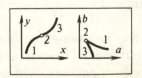

*Fig. 65.* The duality of inflection points and cusps

A curve of parabolic points on the original surface corresponds to a cusp ridge on the dual surface. The special points on this curve (where it is tangent to the curve of inflection of the asymptotics) correspond to swallowtails. The self-intersection curve of a swallowtail consists of planes doubly tangent to the original surface. Consequently, at a point 7) the two points of tangency of the plane with the original surface coalesce, which means that the one-parameter family of doubly tangent planes terminates.

The classes of points on a generic surface appear also in the form of the various singularities of a visible contour. If the direction of projection is generic, then by Whitney's theorem the only singularities are folds and cusps. However, if one has chosen the direction of projection in a special way, one may also obtain certain non-generic projections of a generic surface. It turns out that all such projections reduce locally to the projection of one of the following nine surfaces $z = f(x, y)$ along the $x$-axis:

| type | 1 | 2 | 3, 4 | 5 |
|------|-----|---------|------------------|-----------|
| $f$ | $x^2$ | $x^3 + xy$ | $x^3 \pm xy^2$ | $x^3 + xy^3$ |

| 6 | 7 | 8, 9 |
|------------|--------------------------|------------------------|
| $x^4 + xy$ | $x^4 + x^2 y + x y^2$ | $x^5 + x^3 y \pm xy$ |

(the surfaces are projected onto the $(y, z)$-plane, and the reduction is realized by a change of coordinates of the form $X(x, y, z)$, $Y(y, z)$, $Z(y, z)$).

The visible contours corresponding to these projections are illustrated in Fig. 66.

The correspondence between the classification of projections and the classification of points on a surface is as follows. Type 1 is a projection along a non-asymptotic direction (a Whitney fold).

The projection along an asymptotic direction at a generic point of the hyperbolic domain belongs to type 2. This projection has a Whitney cusp singularity. Under a small perturbation of the direction of projection the singular point only

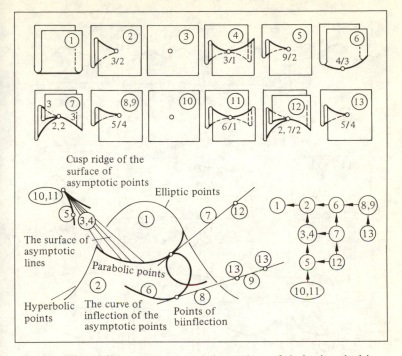

*Fig. 66.* The visible contours and the orders of their singularities
for the typical projections

moves a little on the surface: the new direction turns out to be
asymptotic at a nearby point. So *to see a cusp one has only to
look at a generic surface along an asymptotic direction.*

On movement of the surface or of the observer, at isolated
moments the singularities 3, 4, and 6 will appear.

The projections 6 (and 8 or 9) correspond to the hyperbolic
domain (and to asymptotic tangents of the third and fourth
orders respectively).

On the back of a two-humped camel (see Fig. 43) there is a
curve of inflection of asymptotics. The third-order tangents
passing through its points form a surface. As we walk past the
camel, we intersect this surface twice. At the moment of intersection the visible contour of the back has a singularity of the
type $y^3 = x^4$, and the projection is of type 6.

71

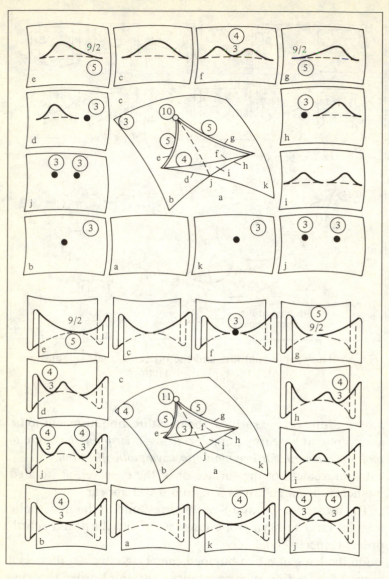

*Fig. 67.* Bifurcations of projections upon deformation of the centre of projection: cases 10–11, $z = x^3 \pm xy^4$

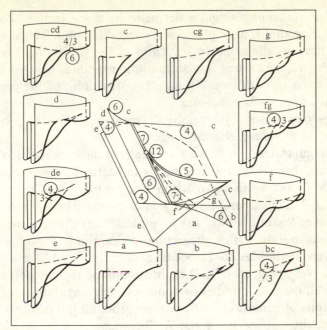

*Fig. 68.* Bifurcations of projections: case 12, $z = x^4 + x^2 y + x y^3$

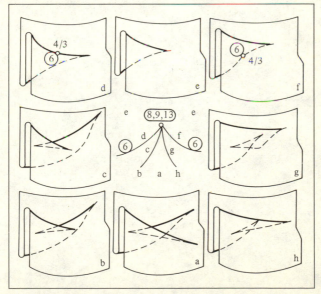

*Fig. 69.* Bifurcations of projections: case 13, $z = x^5 + xy$

The remaining singularities arise upon projection along a direction that is asymptotic at a parabolic point. The simplest of these are the singularities 3 and 4. The projection 3 occurs at the moment when, on descending a mound, we first begin to see its contour (see Fig. 41). The first point of the contour which appears is parabolic.

As one passes a singularity 4, two components of a visible contour come together or separate.

Singularities 5, 7, 8 and 9 are realized only for isolated projecting directions, and one needs to search for them deliberately. (8 and 9 are a projection along a tangent of the fourth order, 7 along a parabolic tangent of the third order, and 5 is a point of "parallelism of the asymptotics at infinitely close parabolic points"). For projections from isolated points four more singularities 10–13 occur: $z = x^3 \pm xy^4$, $z = x^4 + x^2y + xy^3$, $z = x^5 + xy$ (Fig. 66–69). Thus, including the regular point $z = x$ as well the total number of nonequivalent singularities of projections of generic smooth surfaces from all points of the ambient three-space is 14 (O. A. Platonova and O. P. Shcherbak).

# 13 The Problem of Bypassing an Obstacle

Let us consider an obstacle in three-dimensional Euclidean space, bounded by a smooth surface (Fig. 70). It is clear that the shortest path from $x$ to $y$ avoiding the obstacle consists of straight-line segments and segments of geodesics (curves of minimal length) on the surface of the obstacle. The geometry of the shortest paths is greatly affected by the various bendings of the obstacle surface.

Let us assume that the starting point of the paths, $x$, is fixed and let us consider the shortest paths leading from $x$ to all possible points $y$. The paths to points blocked by the obstacle start with straight-line segments that are tangent to the obstacle. The continuations of these paths form a pencil (one-parameter family) of geodesics on the obstacle surface. The following lengths of the paths are again line segments, tangent to the geodesics; they can either terminate at the end point $y$ or once more be tangent to the surface of the obstacle, and so on.

Let us consider the simplest case of a path consisting of an initial and a final straight-line segment with a geodesic segment in between. Neighbouring geodesics of the pencil fill out some domain on the obstacle surface. At each point of this

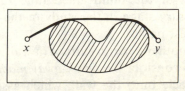

*Fig. 70.* The shortest path avoiding an obstacle

75

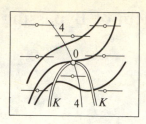

*Fig. 71.* The asymptotic directions and a typical pencil of
geodesics on a surface

domain a geodesic of the pencil has a well-defined direction.
At generic points this direction is not asymptotic. The condition of tangency of a geodesic of the pencil with an asymptotic
direction is one condition on a point of the surface. For a generic surface and pencil this condition is satisfied along some
curve on the surface (depending on the pencil). In Fig. 71 the
asymptotic directions are indicated by horizontal line segments, and the curve of tangency is denoted by the letter $K$;
the geodesics are the heavy curves.

At isolated points (0 in Fig. 71) the curve $K$ itself will have
an asymptotic direction – these are the points of intersection
of $K$ with the curve 4 of inflection of the asymptotics (see
Chap. 12).

In this manner we get a *two-parameter family of paths:* one
parameter indexes the geodesic curve of the pencil, the other
the point of breaking loose of the tangent segment leaving the
surface of the obstacle. Along each path a time function is
defined (reckoned from the starting point $x$). The time needed
to reach the end point $y$ via such a path is not uniquely determined (several such paths may lead to $y$), and besides, not all
of our paths bypass the obstacle. Nevertheless it is clear that
the investigation of the multivalued time function defined
above is a necessary stage in the study of the singularities of
systems of shortest paths.

Let us place another generic surface (a wall) behind the obstacle, and let us consider the *break-away mapping* of the surface of the obstacle onto the wall, which associates to each
point of the obstacle the point of intersection with the wall of

the pencil geodesic tangent breaking away from the obstacle at that point.

When the wall moves away to infinity, the break-away mapping becomes the *Gauss mapping of the pencil:* to each point on the obstacle surface is associated a point on the unit sphere, namely the end of the vector of unit length parallel to the tangent to the geodesic.

*The break-away mapping and the Gauss mapping of the pencil have singularities exactly along the curve where the geodesic directions are asymptotic.* These singularities turn out to be folds at general points and cusps at special points where the direction of the curve is itself asymptotic (O. A. Platonova).

The multivalued time function also has a singularity at points corresponding to an asymptotic breaking away. For a suitable choice of smooth coordinates the time function has the form $T = x - y^{5/2}$ in the neighbourhood of a general point of the singular surface $y = 0$. In other words, if we mark on each ray breaking away the point corresponding to a path of length $T$, then these points form the surface of a front with a cusp ridge, given locally by the equation $x^2 = y^5$ (Fig. 72).

An analogous result is obtained for the plane problem (in this case, the fronts are called involutes and have a singularity of the form $x^2 = y^5$ at points of an inflection tangent (Fig. 73)).

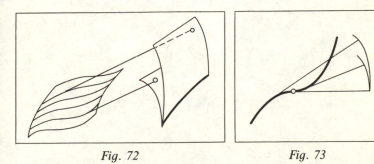

*Fig. 72*          *Fig. 73*

*Fig. 72*. The typical singularity of a front in the problem of bypassing an obstacle: a cusp ridge of degree 5/2

*Fig. 73*. The typical singularity of an involute of a plane curve—a beak of degree 5/2 on an inflection tangent of the curve

The front of the spatial problem at a singular point (a cusp point of the Gauss mapping of the pencil) is given locally by the equations

$$x = u, \ y = v^3 + uv, \ z = (135v^4 + 189uv^2 + 70u^2)v^3,$$

where $(u, v)$ are the parameters and $(x, y, z)$ are curvilinear coordinates in space whose origin lies at a point of the singular asymptotic ray not on the obstacle surface.

# 14 Symplectic and Contact Geometry

Many questions in singularity theory (for instance, the classification of the singularities of caustics and wave fronts, and also the investigation of the various singularities in optimization and variational calculus problems) become understandable only within the framework of the geometry of symplectic and contact manifolds, which is refreshingly unlike the usual geometries of Euclid, Lobachevskij and Riemann.

Let us begin with three examples of singularities of a special type.

## 1. The Gradient Mapping

Suppose that we have a smooth function on Euclidean space. Then the *gradient mapping* is the mapping that associates to each point the gradient of the function at that point. Gradient mappings are a very special class of mappings between spaces of the same dimension.

Singularities of generic gradient mappings are not the same as general singularities of mappings between spaces of the same dimension: there are "fewer" of them, because not every mapping can be realized (up to local diffeomorphisms) as a gradient mapping, but also "more", because phenomena that are not typical for general mappings can be typical for gradient mappings.

## 2. The Normal Mapping

Let us consider the set of all vectors normal to a surface in three-dimensional Euclidean space. Let us associate to each vector its end point (to the vector $p$ based at the point $q$ we associate the point $p+q$). We obtain a mapping of the three-dimensional manifold of normal vectors into three-dimensional space (of an $n$-dimensional manifold to $n$-dimensional space, if we start with a submanifold of any dimension in $n$-dimensional Euclidean space).

This mapping is called the *normal mapping* of the original manifold. The singularities of normal mappings of generic submanifolds constitute a special class of singularities of mappings between spaces of the same dimension. The critical values of a normal mapping form a caustic (the locus of the centres of curvature) of the original submanifold: see Fig. 37, where the original manifold is an ellipse.

### 3. The Gauss Mapping

Let us consider a two-sided surface in three-dimensional Euclidean space. Let us transport the positive unit normal vector from each point of the surface to the origin. The ends of these vectors lie on the unit sphere. The mapping so obtained of the surface into the sphere is called the *Gauss mapping*.

Gauss mappings constitute yet another special class of mappings between manifolds of the same dimension ($n-1$ if we start with a hypersurface in $n$-dimensional space).

It turns out that the *typical singularities of mappings of all three of these classes* (gradient, normal and Gaussian) *are identical:* all three theories are special cases of the general theory of Lagrangian singularities in symplectic geometry.

Symplectic geometry is the geometry of phase space (the space of positions and momenta of classical mechanics). It represents the result of the long development of mechanics, the variational calculus, etc.

In the last century this branch of geometry was called analytical dynamics and Lagrange took pride in having banished all drawings from it. To penetrate into symplectic geometry while bypassing the long historical route, it is simplest to use the axiomatic method, which has, as Bertrand Russell observed, many advantages, similar to the advantages of stealing over honest work.

The essence of this method consists in turning theorems into definitions. The content of a theorem then becomes the *motivation* for the definition, and algebraists usually omit it for the sake of enhancing the authority of their science (to understand an unmotivated definition is impossible, but do many of the passengers in an aeroplane know how and why it is made?).

Pythagoras's *theorem*, a supreme achievement of mathematical culture in its time, is reduced in the contemporary axiomatic exposition of Euclidean geometry to an unobtrusive *definition*: a *Euclidean structure* in a linear space is a symmetric function of a pair of vectors, linear in each argument (a *scalar product*), for which the scalar product of an arbitrary non-zero vector with itself is positive.

The definition of a *symplectic structure* in a linear space is analogous: it is a skew-symmetric function of a pair of vectors, linear in each argument (a *skew-scalar product*), which is non-degenerate (no non-zero vector is skew-orthogonal to every vector, i.e., its skew-scalar product with some vectors is non-zero).

*Example:* Let us take as the skew-scalar product of two vectors on an oriented plane the oriented area of the parallelogram spanned by them (that is, the determinant of the matrix formed of the components of the vectors). This product is a symplectic structure on the plane.

Three-dimensional space (and in general any odd-dimensional space) admits no symplectic structures. It is easy to construct a symplectic structure in four-dimensional (and in general in an even-dimensional) space, by representing the space as a sum of two-dimensional planes: the skew-scalar product splits as the sum of the areas of the projections onto these planes.

All symplectic spaces of the same dimension are isomorphic (as are all Euclidean spaces). We shall call the skew-scalar product of two vectors the "area" of the parallelogram spanned by them.

Every linear subspace of a Euclidean space has an *orthogonal complement*, whose dimension is equal to the codimension of the original subspace.

In symplectic space the *skew-orthogonal complement* of a linear subspace is defined: it consists of all vectors whose skew-scalar products with all vectors of the subspace are zero. The dimension of the skew-orthogonal complement is also equal to the codimension of the original subspace. For example, the skew-orthogonal complement of a line in the plane is this line itself.

A linear subspace which is its own skew-orthogonal complement is called a *Lagrangian subspace*. Its dimension equals one-half the dimension of the original symplectic space.

A *Riemannian structure* on a manifold is given by a choice of a Euclidean structure on the tangent space to the manifold at each point.

In exactly the same way a *symplectic structure on a manifold* is given by the choice of a symplectic structure on each tangent space; however, in contrast to the Riemannian case these structures are not unrestricted, but are interconnected, as is explained below.

A Riemannian structure on a manifold allows us to measure the lengths of curves on it by summing the lengths of small vectors which compose the curve. In exactly the same way, a symplectic structure enables us to measure the "area" of an oriented two-dimensional surface lying in a symplectic manifold (by summing the "areas" of small parallelograms making up the surface). The additional condition connecting the symplectic structures in different tangent spaces is this: the "area" of the whole boundary of any three-dimensional figure should be zero.

One can introduce the structure of a symplectic manifold on any linear symplectic space, by defining the skew-scalar product of vectors based at an arbitrary point to be the skew-scalar product of the vectors obtained from them by parallel translation to the origin. It is easily checked that the compatibility condition is fulfilled here.

There exist many non-isomorphic Riemannian structures in the neighbourhood of a point of the plane or higher-dimensional space (it was to distinguish them that Riemann introduced his curvature).

Unlike Riemannian manifolds, *all symplectic manifolds of a given dimension are isomorphic* (can be mapped onto one another with preservation of "area") *in a neighbourhood of each of their points*. Thus, locally each symplectic manifold is isomorphic to a standard symplectic linear space. In such a space we may introduce coordinates $(p_1, \ldots, p_n, q_1, \ldots, q_n)$ such that the skew-scalar product equals the sum of the oriented areas of the projections onto the planes $(p_1, q_1), \ldots, (p_n, q_n)$.

A submanifold of a symplectic space is called a *Lagrangian manifold* if its tangent space at each point is Lagrangian.

A fibration of a symplectic space into submanifolds is called a *Lagrangian fibration* if all the fibres are Lagrangian manifolds.

Any Lagrangian fibration is locally isomorphic to the standard fibration of the phase space over the configuration space, $(p, q) \rightarrow q$ (the fibres are the spaces of momenta, $q = $ constant). The configuration space (that is, the $q$-space) is called the *base space* of this fibration.

Let us suppose now that in the space of a Lagrangian fibration we are given yet another Lagrangian manifold. Then we get a smooth mapping of this Lagrangian manifold to the base space of the Lagrangian fibration (that is, to the configuration space with coordinates $q_i$): to each point $(p, q)$ of the Lagrangian manifold we associate the point $q$ in the configuration space.

A so-obtained mapping between manifolds of the same dimension $n$ is called a *Lagrange mapping,* and its singularities are called *Lagrangian singularities*.

These form a special class of singularities of smooth mappings between manifolds of the same dimension. For this class a classification theory has been constructed which is analogous to the general theory of singularities.

For $n = 2$ the only generic Lagrangian singularities are folds and cusps, just as for general singularities (however, the Lagrangian cusp has two Lagrangianly inequivalent[1] variants).

---

[1] A *Lagrangian equivalence* between two Lagrangian singularities is a mapping of the first Lagrangian fibration to the second which maps fibres onto fibres, takes the first symplectic structure into the second and carries the first Lagrangian submanifold into the second.

The singularities of Lagrange mappings of three-dimensional generic Lagrangian manifolds are already not all to be encountered among the ordinary generic singularities.

Now we shall show that *gradient, normal and Gaussian singularities are Lagrangian:*

1. Let $F$ be a smooth function of $p$. Then the manifold $q = \partial F / \partial p$ is Lagrangian. Therefore singularities of gradient mappings are Lagrangian.

2. Let us consider a smooth submanifold in Euclidean space. Let us consider the set of all vectors perpendicular to it (at all of its points $q$). The manifold formed by the vectors $p$ attached at the points $p + q$ is Lagrangian. The normal mapping can be considered as a Lagrange mapping from this manifold to the base space, $(p, p+q) \to (p+q)$.

3. Let us consider the manifold of all oriented straight lines in Euclidean space. This manifold is symplectic since it can be considered as the phase space of a point moving on a sphere (the direction of the line determines a point on the sphere, and the point of intersection of the line with the tangent plane of the sphere perpendicular to it determines the value of the momentum).

Let us look at the manifold of positive oriented normals to a surface in our space. This submanifold of the symplectic manifold of oriented straight lines is Lagrangian. The Gauss mapping of the surface can then be considered as a Lagrange mapping (namely, the projection of the submanifold just constructed to the sphere which is the base space of the Lagrangian fibration of the phase space).

Thus the theories of gradient, normal and Gaussian singularities reduce to the theory of Lagrangian singularities.

The symplectic structure on the manifold of oriented straight lines that we encountered just now is not so artificial a formation as it seems at first sight. The point is that the solution set for any variational problem (or, in general, the solution set for a Hamilton equation, with a fixed value of the Hamiltonian) forms a symplectic manifold which is very useful for investigating the properties of these solutions.

Let us consider, for example, the two-parameter family of rays breaking loose from the geodesics of an obstacle surface in three-dimensional space, as shown in Fig. 72. This family turns out to be a two-dimensional Lagrangian sub'manifold' in the four-dimensional space of all rays, but, unlike the Lagrangian submanifolds we encountered earlier, this one itself has singularities, and is called a Lagrangian *variety*. These singularities appear where the rays breaking loose are asymptotic to the obstacle surface. Such rays form a cusp ridge (of the type $x^2 = y^3$) of the Lagrangian variety of all leaving rays.

On this cusp ridge there are special points, in the neighbourhood of which the variety of leaving rays is symplectically diffeomorphic to an open swallowtail (the surface formed in the four-dimensional space of all polynomials $x^5 + ax^3 + bx^2 + cx + d$ by the polynomials which have a triple root).

This space of polynomials has a natural symplectic structure inherited from the $SL(2, \mathbb{R})$-invariant structure of the space of binary forms, and the open swallowtail is a Lagrangian subvariety in the polynomial space. The open swallowtail surface is also encountered in other problems of singularity theory (for example, in the investigation of the sweeping out of caustics by the cusp ridges of moving wave fronts) and appears to be one of the fundamental examples for a future theory of Lagrangian varieties.

In Euclidean and Riemannian geometry there is an extensive theory of external curvature: in addition to the intrinsic properties of a submanifold, determined by its metric, there are also differences in the ways that submanifolds with the same intrinsic geometry are embedded in the surrounding space.

In symplectic geometry, as was recently shown by A. B. Givental', the case is simpler: the intrinsic geometry (the restriction of the symplectic structure to the set of tangent vectors of the submanifold) determines the external geometry. In other words, *submanifolds with the same intrinsic geometry can locally be transformed into one another by a diffeomorphism of the ambient space which preserves the symplectic structure.*

This opens up a new chapter in singularity theory – the investigation of singularities of the disposition of submanifolds in a symplectic space, the importance of which was noted by R. Melrose in recent papers on diffraction. The beginning of a classification of such singularities is obtained, via Givental's theorem, from the results of J. Martinet and his followers on degeneracies of the symplectic structure. For example, a generic two-dimensional submanifold in four-dimensional symplectic space can locally be reduced, by a transformation preserving the symplectic structure, to one of the two normal forms:

$$p_2 = q_2 = 0 \quad \text{or} \quad q_1 = 0, \quad p_2 = p_1^2 .$$

On odd-dimensional manifolds there can be no symplectic structures, but instead there are contact structures. Contact geometry does for optics and the theory of wave propagation what symplectic geometry does for mechanics.

A contact structure on an odd-dimensional manifold is determined by the choice of a hyperplane (subspace of codimension 1) in the tangent space at each point. Two fields of hyperplanes on manifolds of a fixed dimension are locally equivalent (can be transformed into each other by a diffeomorphism), provided they are both generic near the points under consideration.

A *contact structure* is a field of hyperplanes which is generic near each point of an odd-dimensional manifold.

The manifold of all line elements in the plane is a contact manifold. It is three-dimensional. Its contact structure is given as follows: a velocity of motion for an element belongs to the (hyper)plane of the contact field if the velocity of the point of attachment of the element belongs to the element. In exactly the same way we may define a contact structure on the $(2n - 1)$-dimensional manifold of hyperplane elements on an arbitrary $n$-dimensional manifold.

The rôle of the Lagrangian manifolds passes over in the contact case to Legendre manifolds (integral submanifolds of the hyperplane field which have the greatest dimension possible; because of the genericity condition on the hyperplane field this dimension is $m$ in a contact manifold of dimension $2m + 1$).

The singularities of wave fronts, of Legendre transforms and also of hypersurfaces dual to smooth ones are all Legendre singularities. The entire symplectic theory (including, for instance, Givental's theorem) has contact analogues which are extremely useful for the investigation of singularities in variational problems.

Wave propagation in continuous media is described by the *light hypersurface* in the contact space (also called the 'dispersion relation' or the 'zero variety of the principal symbol' in the space of contact elements of space-time).

For waves described by variational principles with hyperbolic Euler-Lagrange equations, the indicated hypersurface generally speaking has singularities.

The variety of singular points of the light hypersurface of a typical variational system has codimension 3 in the contact space. On a three-dimensional space transversal to the singular variety the light hypersurface leaves a trace diffeomorphic to the quadratic cone $u^2 + v^2 = w^2$.

The singularities of light rays and wave fronts are determined by the disposition of the light hypersurface with respect to the contact structure (the rays are just the projections of its characteristics, and the fronts the projections of its Legendre manifolds). An analysis of the typical dispositions reveals the peculiar phenomenon of the *interior scattering* of the waves at the inhomogeneities of the medium.

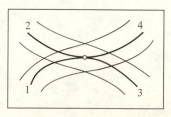

*Fig. 74.* The transformation of waves in a one-dimensional medium

Ordinarily waves of different types (say, longitudinal and transverse waves) are propagated within a medium independently and only on the boundary can one induce the other. But here a transformation of waves comes into effect at interior points of the medium. For example, under the propagation of waves in a one-dimensional nonstationary inhomogeneous medium, scattering is experienced at isolated moments of time by individual rays. The corresponding characteristics in space-time intersect tangentially at one point (Fig. 74).

The curves *1 3* and *2 4* in this drawing are smooth, with a first-order tangency. The characteristics tangent to each other are *1 4* and *2 3*. On a typical wave front moving in three-space the transformation of waves happens at single isolated points.

In recent years symplectic and contact geometry have made their appearance in all areas of mathematics. Just as every lark must grow a crest, so every area of mathematics will ultimately become symplecticized. In mathematics there is a sequence of operations of different levels: functions act on numbers, operators on functions, functors on operators, and so on. Symplectisation is one of the small number of operations of the highest level, which act not on small fry (functions, categories, functors), but on all of mathematics at once. Although several such highest-level operations are presently known (for example, algebraisation, Bourbakisation, complexification, superisation, symplectisation) there is no axiomatic theory whatever for them.

# 15 Complex Singularities

Mathematicians know well that going over to the complex case usually simplifies a problem, rather than making it more complicated. For example, every algebraic equation of degree $n$ has exactly $n$ complex roots, while it is a difficult problem to find the number of real roots.

The reason for this phenomenon is as follows. One complex equation is the same as two real ones. Sets which are given by two real equations (say, curves in space or points in the plane) are said to have codimension two. Sets of codimension two do not divide the ambient space. Therefore, from any point of the space, outside a given set of codimension two, one can reach any other such point by means of a path which avoids this set.

Let us consider some space of complex objects (say, polynomials of a given degree with complex coefficients). The exceptional objects (say, polynomials with multiple roots) are defined by a *complex* equation on the coefficients. Consequently, *the set of exceptional objects has codimension two and does not divide the space of all objects.* For example, the complex swallowtail, formed in the space of complex polynomials $x^4 + ax^2 + bx + c$ by the polynomials with multiple roots, does not divide the space of all such polynomials (which has real dimension 6).

Therefore one may pass from any nonexceptional complex object (for example, from a polynomial without multiple roots) to any other by means of a continuous path, while remaining within the domain of nonexceptional objects (in our example, the domain of polynomials without multiple roots). But if one deforms a nonexceptional object slightly, its topology does not change (for instance, the number of roots of a polynomial without multiple roots does not change if one changes the coefficients sufficiently little). Consequently, all nonexceptional objects of a given class have the same topological invariants (for example, the number of complex roots is the same for all polynomials of a given degree without multiple roots). Thus, all that is needed is to study the topology of a single nonexceptional complex object (e.g. one need only find the number of complex roots of any one equation without multiple roots)[1], in order to know the topology of them all. In contrast, in the real case the set of exceptional objects divides the space of all objects into parts. For example, the ordinary swallowtail (Fig. 34) divides the space of real polynomials $x^4 + ax^2 + bx + c$ into 3 parts: in one of them lie the polynomials with four real roots, in another those with two, and in the third those without any real roots (try to figure out in which part of the space there are so and so many roots!).

As our objects let us now consider curves in the $(x, y)$-plane, given by a condition of the form $f(x, y) = 0$, where $f$ is some arbitrary polynomial of a given degree. For example, if the degree is 2, then the curve will, as a rule, be an ellipse or a hyperbola (all other second-order curves correspond to exceptional, singular cases).

The set of pairs of complex numbers $(x, y)$ which satisfy an equation $f(x, y) = 0$ is called a *complex curve*. From the real point of view this is a two-dimensional surface in four-dimensional space. As a rule (for almost all values of the coefficients of the polynomial $f$) a complex curve is nonsingular. From the previous arguments it follows that all nonsingular complex curves of a given degree are topologically equivalent real surfaces. In order to find out the topology of these surfaces it

---

[1] It is sufficient to take the equation $(x-1)\ldots(x-n) = 0$; very little needs to be added to the arguments adduced above, in order to obtain a completely rigourous proof of the "fundamental theorem of algebra", which says that every equation of degree $n$ has $n$ complex roots.

therefore suffices to study any one of the nonsingular complex curves of a given degree.

The answer turns out to be as follows: the surface is obtained from a sphere by attaching $g = (n-1)(n-2)/2$ handles and removing $n$ points from the surface so formed. For example, a complex line ($n=1$) is a real plane (a sphere with one point removed), a complex circle is a real cylinder (a sphere with two points removed), a complex curve of degree 3 is topologically of the same structure as a torus surface which has been punctured at three places.

The simplest way to convince oneself of this is to obtain a nonsingular curve from a set of $n$ straight lines by means of a slight perturbation. Let us begin, say, with $n$ real lines, situated on the real plane in general position and therefore intersecting each other at $n(n-1)/2$ points (Fig. 75). Each line is given by an inhomogeneous linear equation of the form $l=0$, where $l = ax + by + c$. Let us multiply together the linear functions $l$ corresponding to the $n$ lines. The product vanishes exactly on these $n$ lines. If we replace the curve $f=0$, which splits up into the lines, by a nonsingular curve $f = $ (small number), then we have carried out just the perturbation we need.

If we now go over to complex $x$ and $y$, then each straight line becomes a plane in the real sense, so that the curve $f=0$ turns under complexification into a collection of $n$ real planes. Every pair of such planes in four-dimensional space intersects in a point (because points in fact remain points under complexification). Under the perturbation described above the surface becomes smooth. The smoothing is carried out as follows (topologically): from each of the two intersecting planes

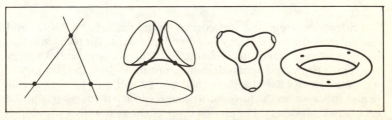

*Fig. 75.* The Riemann surface of a plane algebraic curve

a neighbourhood of the point of intersection is removed, and the two circles bounding the holes which are thereby formed are then glued together (in such a way as to yield an orientable surface).

For example, three spheres intersecting pairwise in single points yield a torus upon smoothing of the three intersection points (Fig. 75). In exactly the same way, $n$ spheres yield a sphere with $(n-1)(n-2)/2$ handles, and so from $n$ planes we get a sphere with this number of handles but with $n$ points removed.

Thus we have solved the problem of determining the topological structure of a nonsingular complex algebraic curve of degree $n$ (the sphere with handles which arose in this construction is called the Riemann surface of the curve)[2].

As for real plane curves of degree $n$, their topological structure is known at present only for curves of low degree (not even the possible arrangements of the branches of a real curve of degree 8 in the plane are known).

Like the theory of curves, the theory of singularities also becomes simpler when we go over to the complex domain; many phenomena which seem quite mysterious from the real point of view acquire a transparent explanation in the complex domain.

For example, let us consider the structure of the simplest critical points of complex functions (i.e., let us consider the complexification of the theory of maxima and minima).

For a real function, critical points are connected with structural changes of level curves or surfaces. For example, the real level curve $x^2+y^2=c$ of the function $f=x^2+y^2$ is empty for $c<0$ and is a circle for $c>0$. For the function $x^2-y^2$ the

---

[2] Incidentally, from the topological properties of the torus (and namely from the fact that any two meridians divide the torus into two parts), it follows that for a mechanical system whose potential energy is a fourth-degree polynomial, oscillations in each of the two potential wells which have the same total energy will also have the same period (because on the toroidal Riemann surface the level sets of the energy in both cases – the phase curves of the two wells – are different meridians).

change of structure is another: the asymptotes of the hyperbola $x^2 - y^2 = c$ are connected in different ways by the branches of the hyperbola, depending on the sign of $c$. In these examples the unique critical value is $c = 0$. The critical level sets are not smooth, whereas the noncritical level sets are smooth manifolds.

In the complex case the axis of function values becomes the plane of the complex variable $c$. The critical values lie at isolated points in this plane and do not divide it into parts. Therefore the level manifolds of $c$ have the same topological structure for all noncritical values $c$. If $c$ changes and passes through a critical value, then no change of structure takes place: true, the level set becomes singular at the moment $c$ passes through the critical value, but then it instantly returns to its original topological type.

In the complex case, instead of *passing through* a critical value, one must *make a circuit around* it (this is a manifestation of the general principle which says that the complex analogue of the real concept "boundary" is the "branched covering").

So on the plane of the complex variable $c$ let us consider a path which turns around some critical value.

To each point on this path corresponds a nonsingular level manifold, $f = c$. As $c$ changes continuously the level manifold changes continuously, while remaining topologically the same.

In other words, we may associate to each point of an initial nonsingular level manifold a nearby point of the nearby level manifold in such a way that we get a one-to-one correspondence, continuous in both directions, between the two level manifolds. Thus we may identify the initial level manifold with the level manifold of any nearby level $c$.

As we change $c$ continuously this identification changes continuously, and finally, when $c$ returns to its original position, we get an identification of the original level manifold with itself. This identification is called the *monodromy*.

So the monodromy is a one-to-one and onto, bicontinuous mapping of the nonsingular level manifold to itself. It turns

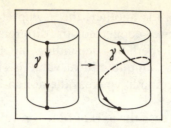

*Fig. 76.* The Dehn twist—the monodromy of the function $x^2 + y^2$

out this mapping is not at all the identity transformation: when $c$ has completed a full turn around the critical value, only the level manifold as a whole for the value $c$ has returned to its place, and not its individual points[3].

In order to understand what happens to the points of a nonsingular level manifold under a monodromy, let us consider the example $f(x, y) = x^2 + y^2$. Let us consider first of all the nonsingular level manifold $x^2 + y^2 = c$, $c \neq 0$. In the real case this equation defines a circle, but we are interested in the "complex circle" – the set of points $(x, y)$ in the plane of two complex variables, such that the sum of the squares of the (complex) coordinates has a given value.

We already know that this surface is topologically a two-dimensional cylinder in four-dimensional space.

It turns out that the monodromy rotates each of the circles composing the cylinder through a different angle, which changes continuously from zero at one base of the cylinder to $2\pi$ at the other. Thus, both edges of the cylinder are left in place pointwise, while the surface is twisted through an entire turn, so that, for example, the generating line of the cylinder is converted into a helix which makes a full turn around the cylinder on its way from one base to the other (Fig. 76).

---

[3] The situation here is exactly the same as on a Möbius band. During a continuous circuit along the axial circle of the Möbius band, we can continuously identify the line segments transverse to it with each other. But when we first return to the original segment, the identification we get of this segment with itself will interchange the ends.

In order to understand why this is so, let us investigate the "complex circle" in greater detail. Its equation can be written in the form $y = \sqrt{c - x^2}$. We can see from this formula that for each (complex) value of $x$ there is a pair of corresponding values of $y$, except for $x = \pm\sqrt{c}$ – for each of these two singular values of $x$ we get a unique value (zero) of $y$.

Consequently, the graph of the complex "two-valued function" $y = \sqrt{c - x^2}$ extends in two sheets over the plane of the complex variable $x$; moreover, the two sheets are joined at only two points. However, if one removes only these two points, one does not succeed in separating the two sheets. In fact, let us have $x$ follow a small loop around one of these points, wrapping around just once. The corresponding value of $y$, changing continuously, returns not to its former value but to a new one. Indeed, from the formula

$$c - x^2 = (\sqrt{c} - x)\,(\sqrt{c} + x)$$

we can see that as $x$ makes a circuit around one of the points $\pm\sqrt{c}$, the argument of one of the factors changes by $2\pi$, while that of the other does not change. Hence the argument of $y$ changes by $\pi$ during this circuit, i.e. $y$ changes sign and transfers from one sheet to the other.

If $x$ makes a twofold circuit around the point $\sqrt{c}$, the quantity $y$ returns to its original value. The points $x = \pm\sqrt{c}$ are called the *branching points* of the function $y = \sqrt{c - x^2}$.

In order to get a better idea of the surface defined by this function, let us join the two branching points by a segment. If the point $x$ wanders about the plane without intersecting this segment, then $y$ will return to its initial value whenever $x$ has described a closed path. Indeed, a single turn around either branching point causes $y$ to change sheets, so a circuit of the entire segment does not change the sign of $y$.

It is clear that our surface $x^2 + y^2 = c$ has the topological structure of a union of two copies of the plane of the complex variable $x$, each slit open along the segment between the two branching points, and joined by gluing the upper edge of the slit in each copy to the lower edge in the other copy. Topolog-

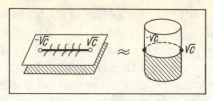

*Fig. 77.* The Riemann surface of the curve $x^2 + y^2 = c$

ically this surface is a cylinder. On this cylinder, the slit appears as the equatorial circle (Fig. 77).

As $c$ approaches the critical value 0, the two branching points draw together. In the limit as $c \to 0$, the segment joining them and the path which goes around the slit segment on the Riemann surface both vanish at the critical point. That is why the equatorial cycle on the cylinder $x^2 + y^2 = c$ is called the *vanishing cycle*.

For $c > 0$ this vanishing cycle is an ordinary real circle. So we have succeeded in understanding the structure of a typical nonsingular level set near a critical point, for a given function value near the critical one. The exact form of the function is not important here, as long as the critical point is nondegenerate. This is so because all nondegenerate critical points of complex functions are topologically locally alike (which is to say, the functions are topologically equivalent near the critical points), in accordance with the general principle explained above (complex degeneracy is a superposition of two real conditions). In particular, the topology of the vanishing cycle is the same for the hyperbolic case ($x^2 - y^2 = c$) as for the elliptic case, $x^2 + y^2 = c$, except that in the hyperbolic case the vanishing cycle lies entirely within the complex domain (and is not visible in the real $(x, y)$-plane).

Now let $c$ circle in a small loop around the critical value. Let us use our analysis of the complex level curve of the function to investigate the monodromy. If we discard a small neighbourhood of the singular point, all level curves (real or complex) for values of $c$ sufficiently close to the critical one can be projected in a one-to-one and bicontinuous way onto

96

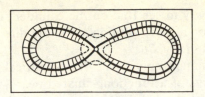

*Fig. 78.* The identification of neighbouring level sets of a function far away from the critical points

the critical level curve outside the aforementioned neighbourhood of the singular point (Fig. 78).

From this it follows that the monodromy, i.e. the identification of the level curves of $c$, which varies continuously along the path run by $c$ during its circuit of the critical value, can be chosen so that outside the aforementioned neighbourhood all points of the level curve return to their places when $c$ completes a full turn.

We must still investigate what happens inside the neighbourhood. For this it is enough to consider the standard function $f = x^2 + y^2$. The part of the complex level curve which falls inside the neighbourhood is topologically a cylinder, both of whose edges lie on the boundary of the neighbourhood. At the same time, this part forms a two-sheeted covering of a domain in the plane of the complex variable $x$, with branching at the points $\pm\sqrt{c}$, as was explained above (Fig. 77).

When $c$ completes a full turn around zero, the segment between the branching points completes a half turn, and as a result we revert to the former branching points, albeit transposed. If we continuously identify with each other the Riemann surfaces arising during $c$'s journey (in such a manner that the edge points remain near to their original positions the whole time), we obtain finally a mapping of the cylinder to itself (the monodromy), organized in the following way.

A segment of a generating line of the cylinder , denoted by the letter $\gamma$ in Fig. 79, 1, subsequently is identified to the curves denoted by this same letter on the intermediate surfaces (2, 3, 4). Finally, we revert to the original cylinder (5), but the curve $\gamma$ has gone over into a new curve having the same end points. It is easy to see that on the cylinder surface this new curve makes one full revolution along the directrix circle, just as is depicted in Fig. 76.

Thus, the monodromy twists the cylindrical part of a complex level curve of a function, situated near a nondegenerate critical point, through exactly one whole turn. The vanishing cycle goes over into itself under this twisting (having been rotated by $\pi$). But other cycles on the level curve are in general transformed into new cycles. Namely, every time some cycle passes through along a generating line of our cylinder (i.e. cuts across the vanishing cycle) the twisting alters the cycle passing through by a copy of the vanishing cycle, so that (up to continuous deformations) the image of the passing cycle under the monodromy can be obtained by adding to the passing cycle as many copies of the vanishing cycle, as the number of times (counted with regard for sign) that the cycle passing through crosses the vanishing cycle. If this number equals zero, then the passing cycle is said to be orthogonal to the vanishing cycle. Such a cycle does not change under the monodromy.

In this way we have deduced (for functions of two variables) the "Picard-Lefschetz formula", which is fundamental in the complex theory of critical points of functions. If we go over to functions of an arbitrary number $n$ of variables, the vanishing cycle becomes a sphere of dimension $n-1$ and the cylinder is replaced by the set of all tangent vectors of this sphere. If the number of variables $n$ is odd, then the monodromy acts on the classes of cycles as reflection in a mirror orthogonal to the vanishing cycle (which itself changes sign under the monodromy).

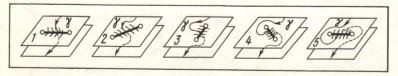

*Fig. 79.* The construction of the monodromy by successive identification of nearby Riemann surfaces

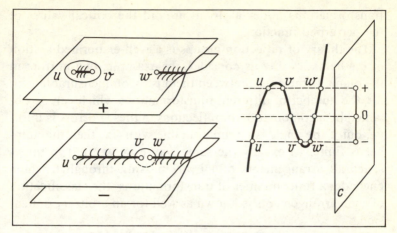

*Fig. 80.* The vanishing cycles of the function $x^3 + y^2$

Complicated (i.e. degenerate) critical points of functions decompose under generic small perturbations into the simplest (i.e. nondegenerate) ones. As the result of a generic small perturbation, several nondegenerate critical values arise, and near each of these a vanishing cycle. Going around any of the critical values defines a monodromy transformation. If from a noncritical original value one approaches any of the critical values along a noncritical path, then one can transport the vanishing cycle of this critical value into the level manifold of the original nonsingular level of the perturbed function. As a result a whole collection of vanishing cycles arises.

For example, a nonsingular complex level curve of the function $x^3 + y^2$ is a torus minus a point. The slightly perturbed function $x^3 - \varepsilon x + y^2$ has two critical values (Fig. 80). Approaching them from the noncritical complex level curve defines two vanishing cycles on this torus: a parallel and a meridian of the torus. In exactly the same way, on the level surface of the function $x^3 + y^2 + z^2$ lie two vanishing spheres intersecting in a single point. The corresponding monodromy transformations are reflections of the space of cycle classes in mirrors orthogonal to the vanishing cycles.

Thus, reflection groups appear in the theory of critical points of functions: they are constituted by the monodromy

99

transformations under a circuit around the critical values of the perturbed function.

The theory of reflection groups is a well-elaborated section of mathematics. Let us consider, for example, two mirrors in the plane. If the angle between them is incommensurable with $2\pi$, the number of different transformations which can be obtained by a combination of reflections in these mirrors is infinite, but if the angle is commensurable with $2\pi$, then the number is finite. In exactly the same way may we find in three-space all arrangements of mirrors passing through 0 which engender a finite number of transformations; the classification of such arrangements is known as well for an arbitrary dimension.

The calculation of the monodromy groups of the simplest degenerate critical points of functions revealed a profound connection between the theory of critical points of functions and the theory of caustics and wave fronts on the one hand and the theory of groups generated by reflections on the other.

The manifestations of this connection sometimes look quite unexpected. Let us consider, for example, the problem of bypassing an obstacle in the plane, bounded by a generic smooth curve with an ordinary inflection point. The time level lines in this problem are the involutes of the curve. These involutes have singularities on the curve (of order 3/2) and on the inflection tangent (of order 5/2). It turns out that the changes in

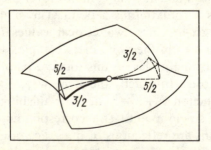

Fig. 81. The discriminant of the symmetry group of the icosahedron—the typical singularity of the graph of the multivalued time function on a surface with boundary

100

structure of the singularities of the involutes as one passes through the inflection point are governed by the symmetry group of the icosahedron. From this one can deduce, for example, that the graph of the (multivalued) time function in a neighbourhood of the inflection point can be reduced by a smooth change of coordinates to a normal form somewhat like the swallowtail. In fact, the normal form is the surface of polynomials $x^5 + ax^4 + bx^2 + c$ with multiple roots (or the surface obtained as the union of the tangents to the curve $(t, t^3, t^5)$, Fig. 81; O. V. Lyashko, O. P. Shcherbak).

# 16 The Mysticism of Catastrophe Theory

The applications of singularity theory to the natural sciences do not exhaust all the directions taken by catastrophe theory: along with concrete investigations of the Zeeman type, one has the more philosophical work of the mathematician René Thom, who first recognized the all-encompassing nature of Whitney's work on singularity theory (and the preceding work of Poincaré and Andronov on bifurcation theory), introduced the term 'catastrophe', and has engaged in a broad propaganda for catastrophe theory.

A qualitative peculiarity of Thom's papers on catastrophe theory is their original style: as one who presenses the direction of future investigations, Thom not only does not have proofs available, but does not even dispose of the precise formulations of his results. Zeeman, an ardent admirer of this style, observes that the meaning of Thom's words becomes clear only after inserting 99 lines of your own between every two of Thom's.

To enable the reader to form his own impression of this style, I shall give a sample from a survey of the perspectives of catastrophe theory done by Thom in 1974.

"On the plane of philosophy properly speaking, of metaphysics, catastrophe theory cannot, to be sure, supply any answer to the great problems which torment mankind. But it favorizes a dialectical, Heraclitian view of the universe, of a world which is the continual theatre of the battle between 'logoi', between archetypes. It is a fundamentally polytheistic outlook to which it leads us: in all things one must learn to recognize the hand of the Gods. And it is perhaps in this as well that it will come across the inevitable limits of its practicality. It will perhaps suffer the same fate as psychoanalysis. There is hardly any doubt that the essence of Freud's discoveries in psychology is true. And yet, the knowledge of these facts has itself been of but very little effectiveness on a practical level (for the treatment of mental disorders, in particular). Just as the hero of the Iliad could go against the will of a God, such as Poseidon, only by invoking the power of an opposed divinity, such as Athena, so shall we be able to restrain the action of an archetype only by opposing to it an antagonistic archetype, in an ambiguous contest of uncertain outcome. The same reasons which permit us to extend our possibilitites of action in some cases will condemn us to impotence in others. One will perhaps be able to demonstrate the inevitable nature of certain catastrophes, such as illness or death. Knowledge will no longer necessarily be a promise of success or of survival; it might just as well mean the certainty of our failure, of our end."

The beautiful results of singularity theory happily do not depend on the dark mysticism of catastrophe theory. But in singularity theory, just as in all mathematics, there is a mysterious element: the astonishing concurrences and ties between objects and theories which at first glance seem far apart.

One example of such a concurrence which remains enigmatic (although partly understood) is the so-called $A$, $D$, $E$-classification. It is encountered in such diverse areas of mathematics as, for example, the theories of critical points of functions, Lie algebras, categories of linear spaces, caustics, wave fronts, regular polyhedra in three-dimensional space and Coxeter crystallographic reflection groups.

Common to all of these cases is the requirement of *simplicity* or *absence of moduli*. Simplicity means the following. Every classification is a partition of some space of objects into classes. An object is called *simple* if all objects near to it belong to a finite set of classes.

*Example 1.* Let us call two sets of lines passing through the point 0 in the plane *equivalent* if one of them can be transformed into the other by a linear transformation $(x, y) \rightarrow (ax + by, cx + dy)$. Any set of three lines is simple (any collection of three different lines is equivalent to the collection $x=0, y=0, x+y=0$). Any set of four lines passing through 0 is not simple (prove it!).

*Example 2.* We shall classify the critical points of (complex) smooth functions, placing functions in the same class if they can be transformed into each other by a smooth (complex) local change of variables. *The list of the simple singularities* (say for functions of three variables) *consists of two infinite series* and *three exceptional singularities:*

$$A_k = x^2 + y^2 + z^{k+1}, \; k \ge 1;$$
$$D_k = x^2 + y^2 z + z^{k-1}, \; k \ge 4;$$
$$E_6 = x^2 + y^3 + z^4,$$
$$E_7 = x^2 + y^3 + y z^3,$$
$$E_8 = x^2 + y^3 + z^5.$$

*Example 3.* A *quiver* is a set of points together with some arrows joining them. If to every point there is associated a linear space (point, straight line, plane ...) and to every arrow a linear map (from the space corresponding to the beginning of the arrow to the one corresponding to the end), then one says that a *representation of the quiver* has been given. Two representations are called *equivalent* if one goes over into the other under suitable linear transformations of the spaces.

In Fig. 82 the quiver on the left is simple, the one on the right is not (see example 1).

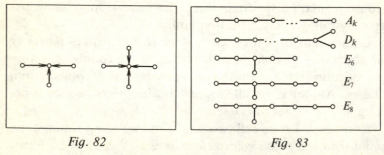

*Fig. 82*        *Fig. 83*

*Fig. 82.* A simple and a nonsimple quiver

*Fig. 83.* The Dynkin diagrams which define simple quivers

As it turns out, *the connected simple quivers are precisely those which are obtained by laying out arrows in an arbitrary way on the Dynkin diagrams depicted in Fig. 83, which make up two infinite series and three exceptional diagrams*.

Simple singularities of caustics and wave fronts also form two infinite series $A_k$ and $D_k$ and three exceptional singularities $E_k$ (the initial members of the series are shown in Figs. 34–35).

The symmetry groups of regular polyhedra in three-dimensional space also make up two infinite series and three exceptions (the exceptions are the symmetry groups of the tetrahedron ($E_6$), of the octahedron ($E_7$), and of the icosahedron ($E_8$), the series are the groups of the regular polygons and the regular dihedra, i.e., of two-sided polygons with the faces painted different or the same colours).

At first glance, functions, quivers, caustics, wave fronts and regular polyhedra have no connection with each other. But in fact, corresponding objects bear the same label not just by chance: for example, from the icosahedron one can construct the function $x^2 + y^3 + z^5$, and from it the diagram $E_8$, and also the caustic and the wave front of the same name.

To easily checked properties of one of a set of associated objects correspond properties of the others which need not be evident at all. Thus the relations between all the *A, D, E*-classifications can be used for the simultaneous study of all simple objects, in spite of the fact that the origin of many of these relations (for example, of the connections between functions and quivers) remains an unexplained manifestation of the mysterious unity of all things.

In the words of the poet:

> The world's made all in one. And its integrity
> The planet does not tire ever to air –
> And lo, the eye is struck by the close kinship
> Of light now shining here, now shining there.
> For certainly there is some sort of kernel,
> From which the light diverges everywhere:
> Into the ripe light of September's bounties,
> Into the flickering wonder of our life.

Interpretations in terms of singularity theory were found in 1983 for all the Coxeter groups generated by reflections in Euclidean spaces, including the noncrystallographic ones like $H_3$ and $H_4$.

The groups $B_k$, $C_k$ and $F_4$ are related to *boundary singularities* of functions (1978). It seems that catastrophe theorists are still unaware of the relation of boundary singularities to the simplest (and most important) cases of the so-called theory of *imperfect bifurcations*. More complicated cases of the latter theory are connected with the Goryunov theory of projections of complete intersections, which is a far-reaching generalization of the theory of boundary singularities. In Goryunov's theory, in particular, the exceptional group $F_4$ turns out to be the progenitor of an entire family of singularities $F_k$, $k \geq 4$.

The geometrical interpretation of the $F_4$ caustic is due to I. G. Shcherbak. Let us consider a surface with boundary in the usual Euclidean 3-space. *The caustic of a surface with boundary* is made up of three surfaces: the focal set of the original surface (formed by its centres of curvature), the focal set of the boundary curve (which is the envelope of its normal planes), and the surface formed by the normals to the original surface at the boundary points. For a generic surface with boundary, there will be some points of the boundary at which the princi-

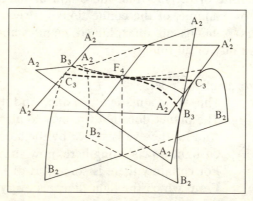

*Fig. 84.* The caustic of the group $F_4$—the typical singularity of the focal set of a surface with boundary

pal curvature direction of the surface is tangent to the boundary. At the focal point along the normal to the surface at such a point, the caustic of the surface will be locally diffeomorphic to the caustic of the group $F_4$ (Fig. 84).

$H_3$, the symmetry group of the icosahedron, is related to the metamorphoses of the involutes of a plane curve at the curve's inflection points. In the corresponding plane obstacle problem the graph of the many-valued time function is diffeomorphic to the variety of singular $H_3$ orbits; it is also diffeomorphic to the union of all the tangent lines of the curve $x = t$, $y = t^3$, $z = t^5$ (O. V. Lyashko, O. P. Shcherbak). In the obstacle problem in three-space the same variety describes the singularity of the front at some points on the obstacle surface.

$H_4$ is the symmetry group of a regular polyhedron with 600 faces in Euclidean 4-space. To construct this polyhedron, one begins with the rotation group of an icosahedron. Under the double covering of SO(3) by SU(2), this rotation group, which is a subgroup of order 60 in SO(3), is covered by the "binary icosahedral group" of order 120. SU(2) is naturally isometric to the 3-sphere $S^3$, and the 120 elements of the binary group form the vertices of the desired regular polyhedron in $\mathbb{R}^4$.

Now let us consider the obstacle problem in 3-space. The graph of the (multivalued) time function is a hypersurface in 4-dimensional space-time. For a generic obstacle problem this hypersurface is at some point locally diffeomorphic to the variety of singular $H_4$ orbits, namely at some point lying on a tangent to a geodesic on the obstacle surface asymptotically tangent to the surface at a parabolic point (O. P. Shcherbak, 1984).

# Appendix
## The Precursors of Catastrophe Theory

At first a thought, to body made
In verse concise by some dear poet,
Like a young maiden lives in shade,
The inattentive world won't know it;
Then, grown more daring, it will be
Adroit now, artful and loquacious,
From every side quite plain to see,
A knowledgeable wife is she
In novels' prose free and audacious;
Become an aging gossip then,
Sustaining saucy exclamations,
In journalistic disputations
She breeds what's long been known to men.

*E. Baratynskij*

Without aspiring to completeness, I shall cite here some of the distinguished work whose authors looked at singularities, bifurcations and catastrophes in generic systems arising in various fields of knowledge.

Caustics are already met in the work of Leonardo da Vinci. Their name was given them by Tschirnhaus.

In 1654 Huygens constructed the theory of evolutes and involutes of plane curves, discovering at the same time the stability of cusp points on caustics and wave fronts (i.e., the stability of the cusp singularities of the corresponding mappings). The metamorphoses of fronts on the plane were investigated by de l'Hôpital (in about 1700) and by Cayley in 1868.

In 1837–1838, Hamilton applied an investigation of critical points of families of functions to the study of singularities of ray systems in geometrical optics, like conical refraction and double refraction.

Jacobi in his lectures on dynamics (1866) investigated the caustics of the system of geodesics of an ellipsoid emanating from one point, and brought to light the stability of the cusp points on the caustics.

The algebraic geometers of the last century knew well the typical singularities of curves (Plücker) and surfaces (Salmon) dual to smooth ones. The swallowtail was described in detail by Kronecker (1878) and found its way into algebra textbooks (Weber, 1898); one can find it in the catalogue of plaster surface models (Brill, 1892) which exist in the geometry rooms of the old universities.

The typical singularities of mappings of surfaces into three-space (the Whitney umbrella, $z^2 = xy^2$, one-half of which is depicted above, in Fig. 31) were investigated by Cayley in 1852. Cayley also studied the geometry of the family of equidistants and the caustic of a triaxial ellipsoid and thereby the 'purse' ('hyperbolic umbilic'), depicted above, in Fig. 39, $c$. He explicitly stated the problem of the topology of the families of level curves of a generic smooth function (1868) and investigated bifurcations in some typical three-parameter families of functions of two variables.

Algebraic analogues of the transversality theorem of singularity theory were systematically used by algebraic geometers, particularly those of the Italian school (Bertini 1882 and others).

Poincaré extensively developed the theory of bifurcations (including cases more complicated than the 'Hopf bifurcation') in his dissertation and in the "New Methods of Celestial Mechanics" (Vol. I, § 37, § 51; Vol. III, Ch. 28 and so on).

Unfortunately, the unsophisticated texts of Poincaré are difficult for mathematicians raised on set theory. Poincaré would have said: "The line divides the plane into two half-planes," where modern mathematicians write simply: "The set of equivalence classes of the complement $\mathbb{R}^2 \backslash \mathbb{R}^1$ of the line $\mathbb{R}^1$ in the plane $\mathbb{R}^2$ defined by the following equivalence relation: two points $A, B \in \mathbb{R}^2 \backslash \mathbb{R}^1$ are considered to be equivalent if the line segment $AB$ connecting them does not intersect the line $\mathbb{R}^1$, consists of two elements" (I am quoting by memory from a schoolbook).

In the book "The Mathematical Heritage of Henri Poincaré", published by the American Mathematical Society, it is even written that Poincaré did not know what a manifold is. In fact, the definition of a (real) smooth manifold is presented in detail in Poincaré's *Analysis Situs*. In modern terms it is this: by a manifold is meant a submanifold of a Euclidean space, considered up to diffeomorphisms.

This simple definition is better than modern axiomatic constructions to the same degree that the definition of a group as a transformation group (considered up to isomorphism), and the definition of an algorithm based on some kind of (universal) Turing machine, are more intelligible than abstract definitions.

Abstract definitions arise in attempts to generalise 'naive' concepts while preserving their basic properties. Now that we know that these attempts do not lead to a real extension of the circle of objects (for manifolds this was established by Whitney, for groups by Cayley, for algorithms by Church), would it not be better to go back to the 'naive' definitions in teaching as well?

Poincaré himself discusses at length the methodic advantages of naive definitions of a circle and a fraction in "Science and Method": it is not possible to master the rule for addition of fractions without cutting up an apple or a pie, if only mentally.

In 1931 A. A. Andronov came forward with an extensive program, which differed from the modern-day program of catastrophe theorists only in that the place of Whitney's singularity theory, which had not yet been created at that time, was taken by the qualitative theory of diferential equations and Poincaré's bifurcation theory. The ideas of structural stability (*grubost'* = roughness), codimension (degree of nonroughness), bifurcation diagrams, the explicit classification of generic bi-

110

furcations and even the investigation of folds and cusps of smooth mappings of a surface onto the plane are explicitly present in the work of A. A. Andronov and his school.

Physicists have always used constructions more or less equivalent to catastrophe theory in the investigation of concrete problems. In thermodynamics these ideas were systematically used by Maxwell and particularly by Gibbs (1873). The metamorphosis of the isotherms of the van der Waals diagram is a typical example of an application of the geometry of the cusp. An analysis of the asymptotics in the neighbourhood of the critical point quickly leads to an understanding that this geometry is not dependent on the exact form of the state equation – a fact which has been well known since Maxwell's time and which is mentioned in most thermodynamics textbooks (for example, in that of Landau and Lifschitz). Maxwell's proposal to draw the horizontal part of the isotherm so that the areas of the holes above and below it are equal signifies a transition from one of the two competing minima of the potential to the other at the moment the second becomes lower. The corresponding bifurcation diagram in catastrophe theory is called the *Maxwell stratum*. Gibb's 'phase rule' furnishes topological constraints on the structure of this and similar bifurcation diagrams (bringing to the open the necessity of strictly proving such facts is the merit of mathematicians of a later period). Gibbs also explicitly pointed out the connection of thermodynamics with the geometry of the contact structure.

Geological applications of the analysis of singularities were indicated by Schreinemakers (1917).

In Semenov's theory of 'thermal explosion' (1929) and in the work of his followers on the theory of combustion the transformations of the stationary modes of behaviour upon change of parameter were explicitly studied, which led to the necessity of investigating folds, and cusps, and more complicated situations as well. In particular, in a 1940 paper by Ya. B. Zel'dovich the phenomena which happen under Morse surgery of the equilibrium curve on the plane of the phase variable and the parameter are analyzed (the birth of new islets or their fusion with the main curve). In the contemporary mathematical theory of relaxation oscillations the analogous analysis has been carried out only within the last few years.

The analysis of a wave field near a caustic and its singularities led Airy and Pearcey to oscillating integrals whose phase gives a normal form for the fold and the cusp respectively. In connection with this it is worth noting that the asymptotics of a field near a boundary found by M. A. Leontovich and V. A. Fok have still not been assimilated by catastrophe theory.

In the theory of elasticity, Koiter in 1945 discovered a semicubic singularity in the dependence of the limiting load upon how far off-center it is applied in the problem of the snapping of an arch. Specialists in elasticity theory have used the cusp geometry in choosing a testing program for elastic constructions in which snapping does not occur despite high loads.

In the investigations, calculations were usually carried through without a general theory, by throwing away in the right fashion some terms of the Taylor series and retaining the other 'most important' ones. Of the physicists who particularly systematically applied catastrophe theory before it arose, we should especially single out L. D. Landau. In his hands the art of throwing away 'inessential' terms of the Taylor series, preserving smaller-sized 'physically important' terms, yielded many results included in catastrophe theory.

Thus, in a 1943 paper on the onset of turbulence, Landau by this method directly writes out the equation of the 'Hopf bifurcation' for the squared amplitude of an oscillation losing stability. The theory of second-order phase transitions following Landau reduces to an analysis of the bifurcations of critical points of symmetric functions. The Landau curves in the theory of Feynman integrals depending on parameters, with their stable cusp points, are included among the fundamental bifurcation diagrams of contemporary catastrophe theory.

Of course, the modern general theory permits one to investigate more complicated singularities with less expenditure of effort. However, the greatest practical value belongs in most cases precisely to investigations of the simplest and most frequently encountered singularities: the expenditure of effort in surmounting the technical difficulties standing in the way of an investigation of the more complicated cases is not always justified by the practical value of the results obtained. On the contrary, the fundamental work of the precursors of catastrophe theory (those mentioned above as well as many others) retains all its significance even now that its mathematical structure has been completely cleared up by singularity and bifurcation theory.

# Conclusion

The newspapers bring tidings of ever new catastrophes. Earthquakes, floods, explosions, wars, epidemics are all around us, and in addition the threat of the most terrible of catastrophes – the nuclear catastrophe – hangs over the whole globe. It's time to put a ban on atomic civil war.

Mathematical catastrophe theory by itself cannot ward off catastrophes, just as the multiplication table, for all its usefulness for bookkeeping calculations, is neither a rescue from theft by individuals nor from an unwise organisation of the economy as a whole.

The mathematical catastrophe models indicate, however, some common traits of the most diverse phenomena of jump change in a system's behaviour in response to a smooth change of the external conditions. For example, a stable steady-state mode of behaviour (let us say, the working mode of a reactor, or an ecological or economic mode of behaviour) usually perishes either by colliding with an unstable mode (where at the moment of collision the speed of convergence is infinitely large), or in consequence of the (again infinitely rapid) growth of self-sustaining oscillations. This explains why it is so hard to fight a catastrophe once its symptoms have already made themselves noticeable: the speed of its approach grows unboundedly in proportion to one's coming nearer to the catastrophe.

A catastrophic loss of stability may be the result of optimisation and intensification. For example, for the simplest model of fishing

$$\dot{x} = x - x^2 - c$$

114

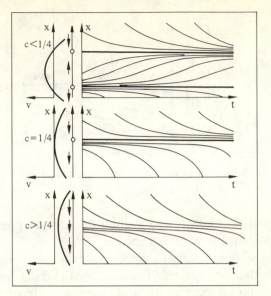

*Fig. 85.* Catastrophic loss of stability upon optimisation in the simplest model of fishing taking competition for food into account

optimisation (maximisation) of the catch quota $c = 1/4$ leads to instability of the steady-state mode of behaviour (Fig. 85) and to a catastrophe – the annihilation of the population by small random oscillations.

Stability will not be lost if one introduces *feedback*: for the rigid plan $c$ one substitutes a quantity proportional to the actually existing resources (the harvest, the population, ...). In the model with feedback (Fig. 86)

$$\dot{x} = x - x^2 - kx$$

the optimal value for the coefficient $k$ is $1/2$. With this choice a many-years' average catch of $kx_0 = 1/4$ will establish itself. This is the same catch as the maximal rigid catch plan (a greater productivity is not possible in this system).

But while under the maximal rigid plan the system loses stability and self-destructs, the introduction of feedback stabilises it and, for example, small changes of the coefficient $k$ (or

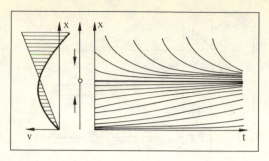

*Fig. 86.* Stabilisation upon replacing the rigid plan by feedback

other fortuities) lead only to a small decrease in productivity, but by no means to a catastrophe.

Control without feedback always leads to catastrophes: it is important that persons and organisations making responsible decisions should personally and materially depend on the consequences of these decisions.

Aggressors who unleash wars or interethnic hostility usually reckon that they will not bear personal responsibility for the consequences, and the fear of personal nuclear or prison-camp annihilation serves as an important deterring factor.

Scholars who were investigating models for the arms race predicted already in the sixties that the introduction of multiple warheads would entail a loss of stability of the strategic balance. They also predicted that if by a diplomatic route we should manage to get past this dangerous period safely, then the subsequent rise in cost of weaponry would stabilise the situation and stability could be restored.

The present *perestroika* can to a large extent be explained by the fact that at least some mechanisms of feedback have begun to act (the fear of personal annihilation).

The difficulty of the problem of *perestroika* is connected with its nonlinearity. Habitual management methods, under which results are proportional to efforts, do not work here, and it is necessary to develop a specifically nonlinear intuition founded on the at times paradoxical conclusions of the nonlinear theory.

116

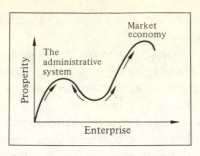

*Fig. 87. Perestroika* from the point of view of the theory of
metamorphoses (perestroikas)

The mathematical theory of metamorphoses (in Russian: *perestroikas*) was created long before the present *perestroika*. Here are some of the simplest qualitative deductions out of this theory with regard to a nonlinear system in an established stable state acknowledged to be bad, inasmuch as there is a better, preferred stable state of the system within visibility (Fig. 87).

1. Gradual motion in the direction of the better state at once leads to a worsening. The speed of deterioration under uniform motion toward the better state is increasing.

2. As one moves from the worse state to the better the resistance of the system to a change of its state grows.

3. The maximum of the resistance is attained sooner than the worst state through which it is necessary to pass in order to reach the better state. After passing the maximum of the resistance the state continues to grow worse.

4. As one approaches the worst state on the path to *perestroika* the resistance from a certain moment onward begins to decrease, and as soon as the worst state has been passed, not only does resistance completely vanish, but the system starts to be attracted toward the better state.

5. The magnitude of the worsening necessary for a transition to the better state is comparable to the final improvement and increases in proportion to the perfection of the system. A weakly developed system can go over to the better state almost without a prior worsening, whereas a well-developed system,

by virtue of its stability, is not capable of such a gradual, continuous improvement.

6. If one can manage to move the system out of the bad stable state all at once, by a jump and not continuously, near enough to the good state, then subsequently it will evolve toward the good state all by itself.

These objective laws for the functioning of nonlinear systems cannot be left out of consideration. Above we have formulated only the simplest qualitative conclusions. The theory also furnishes quantitative models, but the qualitative deductions are the more important and at the same time the more trustworthy: they do not depend much on the details of the functioning of the system, whose mechanism and numerical parameters may be insufficiently known.

Napoleon criticised Laplace for his 'attempt to introduce the spirit of infinitesimals into government'. The mathematical theory of metamorphoses (or perestroikas) is that part of contemporary infinitesimal analysis without which a conscious management of complicated and poorly known nonlinear systems is practically impossible.

However, a special mathematical theory is not required in order to understand that a disregard for the laws of nature and society (be it the law of gravity, the law of value or the necessity of feedback), a decline in the competence of experts and the absence of personal responsibility for decisions taken leads sooner or later to a catastrophe.

# Problems

## To Chapter 1

*(here and in the following the variable z is complex, x and y are real)*

**1.** Find the critical points and the critical values of the mappings $z \mapsto z^2$, $z \mapsto z^2 + \varepsilon \bar{z}$.

**2.** Find the critical points and the critical values of the mappings $(x, y) \mapsto (x^2 + ay, y^2 + bx)$.

**3.** Investigate the bifurcations of the singular points of the differential equation $\dot{x} = -x^3 + x + a$ under a change of the parameter $a$.

**4.** Investigate the bifurcations of the singular points in the system of differential equations $\dot{z} = \varepsilon z - z^2 \bar{z} + A \bar{z}^3$, where $A$ is a fixed complex number and the complex number $\varepsilon$ makes a circuit around zero.

**5.** How many topologically different real fifth-degree polynomials $x^5 + \cdots$ are there with four different real critical values? Two polynomials are topologically the same if one can be transformed into the other by a continuous and orientation-preserving change of the dependent and independent real variables.

**6.** Let us denote by $a_n$ the number of types of polynomials $x^{n+1} + \cdots$ with $n$ different critical values (so that the answer to the preceding problem would be denoted by $a_4$), and let us construct the function $p(t) = \sum a_n t^n / n!$. Show that $p(t) = \sec t + \tan t$ (so that the $a_n$ can be expressed via the Bernoulli numbers for odd $n$ and via the Euler numbers for even $n$).

**7.** In the space of polynomials $x^5 + \cdots$ let us consider the region formed by the polynomials with four different real critical values. How many connected components does this region have?

**8.** Let us suppose that the second differential of a smooth function of two variables at a critical point is positive definite. Prove that after a suitable smooth change of the dependent variable $u$ and the independent variables $(x, y)$ the function will reduce to the form $u = x^2 + y^2$.

**9.** Let us suppose that the second differential of a smooth function of $n$ variables at a critical point is a nondegenerate quadratic form. Prove that after a suitable smooth change of the dependent variable $u$

119

and the $n$ independent variables $(x, y)$ the function will reduce to the form $u = x_1^2 + \cdots + x_k^2 - y_1^2 - \cdots - y_l^2$, $k + l = n$.

**10.** Prove that at a critical point of an analytic function of two variables, as a rule, six (complex) inflection points of the level curve disappear.


## To Chapter 2

**11.** How many cusp points does the map $z \mapsto z^2 + \varepsilon \bar{z}$ have?

**12.** Does the map $(x, y) \mapsto (x^2 + ay, y^2 + bx)$ have cusp points?

**13.** Prove that the number of cusp points of a (generic) mapping of the sphere to the plane is even.

**14.** Let a function be given on the sphere whose integral over the sphere is equal to zero and for which zero is not a critical value. Does there exist a smooth mapping of the sphere to the plane, all of whose singularities are folds and which has the given function as its Jacobian determinant?

**15.** Prove that a mapping of the sphere to the plane, all of whose critical points are folds and cusps, can have as its curve of critical points an arbitrary (nonempty) smooth curve on the sphere.

**16.** Let us suppose that all the critical points of a smooth mapping of the sphere to the plane are folds and cusps, and that the number of domains on the sphere where the Jacobian determinant of the mapping is positive is $a$ and the number of domains where it is negative is $b$. Prove that the number of cusps is at least $2|a - b|$.

**17.** Let us associate to each vector normal to an ellipse its endpoint. Prove that the so constructed mapping of the cylinder to the plane has four cusp points.

**18.** If in problem 17 we replace the ellipse by a generic non-self-intersecting curve, then the number of cusp points of the corresponding mapping of the cylinder to the plane is no less than four.


## To Chapter 3

**19.** Let us consider on an ellipse the distance function from the points of the ellipse to a fixed point of the plane. The critical points of these functions form a surface in a three-dimensional manifold – the direct product of the ellipse with its complement in the plane. How many cusps does the projection of this surface onto the plane have? What does the set of critical values of the projection look like?

**20.** In the space of functions on a circle let us consider the set of all functions which have multiple critical values. Does this hypersurface lie one-sidedly or two-sidedly in the space of functions (i.e., can it be provided with a transversal direction which changes continuously all the way up to the self-intersection points and boundary points)?

## To Chapter 4

**21.** Let us consider a parabolic cylinder resting by one of its generating straight lines on a horizontal plane. For which positions of the centre of gravity of the cylinder above the point of tangency is the equilibrium position stable, and for which is it not? Investigate the singularities of the boundary of the region of stability.

**22.** Draw the graph of the function

$$f(u, v) = \min_x (x^4 + ux^2 + vx).$$

## To Chapter 5

**23.** For the system $\dot{x} = x(a + bx + cy)$, $\dot{y} = y(d + ex + fy)$, at which values of the parameters does the equilibrium for which $xy \neq 0$ lose stability? What do the phase curves look like for these values of the parameters?

**24.** Let us consider a vector field on the line depending smoothly on one parameter. Prove that by a smooth transformation of the parameter and a smooth coordinate change on the line, depending smoothly on the parameter, a generic such field can be reduced (in the neighbourhood of a bifurcating singular point) to the field which defines the evolutionary system $\dot{x} = x^2 + a + f(a)x^3$, where $f$ is a smooth function and $a$ is the parameter (in the analytic case all transformations can be made analytic).

**25.** Investigate the equilibrium surface of the two-parameter family of equations $\dot{x} = -x^3 + ax + b$ and the singularities of its projection onto the parameter plane. Which part of the equilibrium surface corresponds to stable equilibria? Investigate the behaviour of a phase point under a slow change of the parameters $a(t)$, $b(t)$.

**26.** Construct a one-parameter family of vector fields on the line corresponding to the bifurcations of Fig. 13.

## To Chapter 6

**27.** When the real parameter $a$ passes through zero, does the equilibrium position of the system $\dot{z} = (i\omega + a)z + Cz|z|^2$ lose stability in the mild or the hard way? Compare the result with Fig. 16.

**28.** Give the bifurcation of Fig. 21 by means of formulas (the components of the field are polynomials of degree 5).

**29.** Investigate the loss of stability of the cycle $z = 0$, $|w| = 1$ of the system

$$\begin{cases} \dot{z} = (a - 1 + \frac{i}{2})z + (a + 1)\bar{z}w \pm \bar{w}(z + \bar{z}w)^3, \\ \dot{w} = iw + w(1 - |w|^2) \end{cases}$$

as the parameter $a$ passes through zero. Find approximately the double cycle branching off and investigate its stability. Compare the results with Fig. 22.

**30.** Investigate the bifurcations of the phase portrait of the system describing $p/q$ resonance, $q \geq 5$

$$\dot{z} = \varepsilon z + z|z|^2 A(|z|^2) + \bar{z}^{q-1}$$

when the small complex number $\varepsilon$ makes a circuit around zero ($A$ is a generic complex function). Compare the results with Fig. 23.

**31.** Investigate the bifurcations of the phase portrait of the system describing $1:3$ resonance, $\dot{z} = \varepsilon z + Az|z|^2 + \bar{z}^2$, when the complex parameter $\varepsilon$ makes a circuit around zero ($A$ is a generic complex number).

**32.** Investigate the bifurcations of the phase portrait of the system describing $1:4$ resonance, $\dot{z} = \varepsilon z + Az|z|^2 + \bar{z}^3$, when the complex parameter $\varepsilon$ makes a circuit around zero (on the plane of the complex variable $A$, forty-eight regions differing in their sequences of bifurcations are known, but it has not been shown even that the number of different stable sequences is finite).

**33.** Investigate the delay of loss of stability in the system $\dot{z} = (i + a)z - z|z|^2 + b$, under a slow change of the parameters $a = \varepsilon t$, $b = c\varepsilon t$.

## To Chapter 7

**34.** Find the stability boundary of the family of equations $\dddot{x} + a\dot{x} + bx = 0$ on the plane of the real parameters $(a, b)$.

**35.** Prove that the stability boundary of the family of equations $\dddot{x} + a\ddot{x} + b\dot{x} + cx = 0$ is diffeomorphic to the surface $w^2 = u^2 v^2$, $u \geq 0$, $v \geq 0$.

**36.** Prove that the stability boundary of the family of equations $\dddot{z} + A\dot{z} + Bz = 0$ in the three-dimensional space $\text{Im} A = 2$ is diffeomorphic to the surface $w^2 = uv^2$, $u \geq 0$, $v \geq 0$.

**37.** Find the number of types of singularities of the stability boundary of a generic family of linear multidimensional systems depending on four parameters.

**38.** Investigate the singularities of the caustic (the envelope of the family of normals) of a triaxial ellipsoid.

**39.** Investigate the singularities of the caustic which is the envelope of the family of geodesics on an ellipsoid emanating from one point.

**40.** Prove that for an arbitrary generic Riemannian metric on the sphere, the caustic which is the envelope of the family of geodesics going out from one point has at least four cusp points.

**41.** Prove that the union of the tangent lines to the curve $\{(t^2, t^3, t^4)\}$ is diffeomorphic to the set of polynomials $x^4 + ax^2 + bx + c$ which have multiple real roots.

**42.** Prove that a smooth function $f(a, b, c)$, whose derivative with respect to $a$ at the origin is different from zero, can be reduced in a neighbourhood of the origin to the form $\pm a + \text{const}$ by a smooth change of coordinates preserving the swallowtail of the preceding problem.

**43.** Prove that a smooth vector field, whose vector at the origin has a non-zero $c$ component, can be reduced in a neighbourhood of the origin to the field $\pm \partial/\partial c$ (which gives the system $\dot{a} = 0, \dot{b} = 0, \dot{c} = \pm 1$) by a smooth change of coordinates preserving the swallowtail of the two preceding problems.

**44.** Let a big caustic in three-dimensional space-time be formed by those values of the parameter $q = (q_1, q_2, q_3)$, for which the function $x^4 + q_1 x^2 + q_2 x$ has degenerate critical points. Draw the metamorphoses of the momentary caustics obtained upon intersecting the big caustic with the isochrones, for the time function $t = q_1 + q_3^2$.

**45.** Prove that a generic time function can be reduced in the neighbourhood of each point of the big caustic of the preceding problem either to the form $t = q_3 + \text{const}$ or to the form $t = \pm q_1 \pm q_3^2 + \text{const}$ by a diffeomorphism of space-time preserving this big caustic.

**46.** Let a big caustic in four-dimensional space-time be formed by those values of the parameter $q = (q_1, q_2, q_3, q_4)$, for which the function $x^4 + q_1 x^2 + q_2 x$ has degenerate critical points. Investigate the metamorphoses of the momentary caustics obtained upon intersecting the big caustic with the isochrones, for the time function $t = q_1 \pm q_3^2 \pm q_4^2$.

**47.** Draw the surface formed by those values of the parameter $q$, for which the function $x^2 y \pm y^3 + q_1 y^2 + q_2 y + q_3 x$ has degenerate critical points.

**48.** Let a big caustic in four-dimensional space-time be formed by those values of the parameter $q$, for which the function $x^2 y + y^4 + q_1 y^3 + q_2 y^2 + q_3 y + q_4 x$ has degenerate critical points. Investigate the metamorphoses of the momentary caustics obtained upon intersecting the big caustic with the isochrones of different generic time functions.

**49.** Draw the image of the $(u, v)$-plane and of its division into lines $u = $ const (or into curves $t = $ const, where $\partial t / \partial u \neq 0$), under the mapping $(u, v) \mapsto (u^2, v, uv)$ into three-space. Compare the answer with Fig. 46 and with Fig. 31.

**50.** Draw the image of a generic surface with a semicubical cusp ridge under the fold mapping of three-space $(u, v, w) \mapsto (u, v, w^2)$ (under the assumption that the tangent plane of the surface at a point of transversal intersection of the cusp ridge with the plane of critical points $w = 0$ does not contain the direction of the $w$ axis). Compare the answer with Fig. 46.

**51.** Draw the surface $y^2 = z^3 x^2$ and compare the answer with Fig. 46 and with the preceding problem.

**52.** Draw the union of the tangents to the curve $\{(t, t^2, t^4)\}$ and compare with the preceding problems.

**53.** Show that the union of the tangents to a generic space curve is locally diffeomorphic to the surface $y^2 = z^3 x^2$ in the neighbourhood of each point where the torsion of the curve vanishes.


## To Chapter 9


**54.** Define a density for a dustlike gravitating one-dimensional medium on a closed curve in the phase plane so that under the motion of particles this curve and this density will be preserved (hint: the curve is $q^2 + p^2 + |p|^3 = 4/27$).

**55.** Prove that as a one-dimensional flow, initially defining a smooth velocity field, of a dustlike medium sweeps past, above a cluster with a root singularity of the density $(a(x, t)x^{-1/2}\theta(x) + b(x, t)$, where $a$ and $b$ are given smooth functions, $a \neq 0$, $\theta(x) = 0$ for $x < 0$, 1 for $x > 0$) the velocity field acquires a weak singularity of the form $c(x, t)x^{3/2}\theta(x)$; by a smooth change of variables one can reduce $c$ to unity.

**56.** Let us consider $N$ particles in the unit cube and surround each of them by a ball of radius $r$. For which minimal $r$ do these balls form a connected chain of unit diameter? Show that the radius decreases like $C/N$ for distributions of the particles along curves, like $C/N^{1/2}$ for distributions along surfaces, like $C/N^{1/3}$ for spatial distributions (the 'dimension', calculated by this method, of the large-scale distribution of galaxies turns out to lie between 1 and 2).

**57.** Draw the nonsmoothness locus of the function

$$F(y) = \min(\min_{x}(x^4 + y_1 x^2 + y_2), y_3)$$

and compare with Fig. 53.

**58.** Draw the metamorphoses of the nonsmoothness curves of the function $F(y_1, y_2, y_3) = \min(y_1, y_2, y_1 + y_2)$, defined in three-dimensional space-time, on the isochrones $t = \text{const}$ for the time function $t = y_1 + y_2 \pm y_3^2$, and compare with Fig. 53.

**59.** Prove that the singularities of generic level surfaces of maximum functions of typical $n$-parameter families of functions are the same ones as the singularities of graphs of maximum functions of generic $n - 1$-parameter families (where the sets for smaller values correspond to the regions lying above the graphs). In this situation 'good' parameter values are those at which the maximum function is less than a fixed constant (and 'good' values of the constant are those which are bigger than the maximum).

**60.** Let us consider the equation $\ddot{x} + k\dot{x} \pm x = 0$.
Determine which values of $k$ correspond to folded foci, which to folded nodes and which to folded saddles on the $(x, E = x^2 + \dot{x}^2)$-plane.

**61.** Find a surface whose asymptotic lines locally form the system of integral curves of a folded focus (node, saddle).

**62.** Prove that the integral curves of the folded saddle which correspond to the separatrices lying on one side of the fold approach the singular point from opposite sides, but the integral curves of the folded node which correspond to the distinguished phase curves of the node lying on one side of the fold approach the singular point from the same side.

**63.** Let us consider a $k$-parameter family of smooth hypersurfaces in an $n$-dimensional linear space equipped with a projection onto an $n - 1$-dimensional subspace. How nonsmooth can the visible contour turn out to be, if the surface being projected is convex and the family is generic?

**64.** Find the number of moduli of the singularities of the convex hulls of typical smooth surfaces in four-space and of typical smooth submanifolds of dimension 3 in five-space.

**65.** The plane curve dual to the curve $y = x^2 + x^{5/2}$ is diffeomorphic to the original curve, but the dual to the diffeomorphic curve $y = x^{5/2}$ is not.

**66.** The curve dual to a typical curve with a singularity of degree $5/2$ also has such a singularity.

**67.** The number of (complex) singular points of type 7 (see Fig. 64) on a typical algebraic surface of sufficiently high degree $d$ is equal to $2d(d-2)(11d-24)$, and the number of points of type 5 is $5d(d-4)(7d-12)$.

**68.** When the level surface of a typical function of three variables approaches a critical level surface, 24 (complex) points of type 7 (Fig. 64) disappear at the critical point.

*To Chapter 13*

**69.** The involute of a plane curve which passes through an ordinary inflection point of the curve has a singularity of type $5/3$ there.

**70.** Draw the involutes of the cubic parabola $y = x^3$.

**71.** Draw the graph of the (triple-valued) time function near a cubical inflection point of a curve on the plane bounding an obstacle.

**72.** Draw the surface formed in the three-dimensional space of line elements on the plane by the elements tangent to the involutes of a plane curve near a (cubical) inflection point of this curve. Which singularities does this surface have, and which ones has its projection onto the plane (associating to each line element its point of attachment)?

**73.** On an obstacle surface let us consider a function equal to the sum of the distance to a goal (along a straight line) and the distance to some starting point along the obstacle surface. Prove that the multiplicities of the critical points of this function are even.

**74.** The equation $C = \int_0^x (t^3 + At + B)^2 \, dt$, $x^3 + Ax + B = 0$, defines a surface in the space with coordinates $(A, B, C)$. Draw this surface and investigate its singularities (it is locally diffeomorphic to the front in the spatial problem of going around an obstacle at a point corresponding to a cusp of the Gauss mapping of the pencil, and its cusp ridge of degree $5/2$ has a semicubical cusp point at the origin).

*To Chapter 14*

**75.** How many symplectically inequivalent planes of dimension $k$ has a symplectic space of greater dimension? Prove that their number is equal to the integer part of $k/2$.

**76.** By a complete flag in a linear space one means a set of subspaces of all dimensions consecutively embedded in one another. How many symplectically inequivalent complete flags does a symplectic space of dimension $2n$ have? Prove that their number is equal to $(2n-1)!! = 1 \cdot 3 \cdot 5 \cdots (2n-1)$.

**77.** On the space of homogeneous polynomials of an odd degree in two variables there is a symplectic structure invariant with respect to the natural action of the group of area-preserving linear transformations of the plane; this structure is unique (up to a non-zero factor). Find an explicit expression for it in terms of the coefficients of the polynomials.

**78.** On each fibre of a Lagrangian fibration there exists a natural local affine structure (a selected class of coordinate systems in which Lagrangian equivalences give affine transformations).

**79.** Prove that the graph of the Legendre transform of a smooth function is a front (the image of a Legendre mapping of a smooth Legendre manifold).

**80.** The feet of the perpendiculars dropped from the origin to the tangent planes of a surface not containing the origin in Euclidean space form a surface called the pedal (and the original surface is called the negative pedal for its pedal). Prove that the singularities of the pedals of smooth surfaces are Legendrian (i.e., that the pedal is diffeomorphic to the front of a Legendre mapping).

**81.** (continuation). Prove that the singularities of the negative pedals of smooth surfaces are Legendrian. Draw the negative pedals of an ellipse in the plane and of an ellipsoid in three-space.

**82.** Of which Legendre mapping is the equidistant of a smooth surface in Euclidean space the front?

**83.** Of which Legendre mapping is the graph of the (many-valued) distance function to a given smooth surface in Euclidean space the front?

**84.** Prove that on the fibres of a Legendre fibration there exist natural structures of locally projective spaces (so that Legendre equivalences, i.e., diffeomorphisms preserving the contact structure and the Legendre fibration structure, give projective transformations on the fibres).

## To Chapter 15

**85.** Let us extend the action of the group generated by the reflections of the plane in two mirrors forming an angle of $\pi/q$ to the complex plane. Prove that the orbit variety is itself homeomorphic to the complex plane, and the variety of nonregular orbits (the orbits of the points of the mirrors) is homeomorphic to the curve $z^2 = w^q$ on the plane of two complex variables.

**86.** Let us extend the action of the group generated by the reflections of the three-space $x_1 + x_2 + x_3 + x_4 = 0$ in the diagonal planes $x_i = x_j$ to the complex space. Prove that the orbit variety is complex three-space and the variety of nonregular orbits is the complex swallow-tail.

**87.** Fig. 81 depicts the real part of the variety of nonregular orbits of the action of the icosahedral symmetry group on complex space. Where are the real orbits situated?

**88.** The monodromy group transformations determined by the function $x^3 - \varepsilon x + y^2$ act on the torus minus a point. Prove that an arbitrary closed non-self-intersecting curve on the torus minus a point, not contractible on the torus, can be taken over into any other such curve by a transformation in the monodromy group.

**89.** How many handles does a complex nonsingular level curve of the function $z^n + w^2$ have? Prove that their number is equal to $g$ if $n = 2g + 1$ or $2g + 2$.


## To Chapter 16


**90.** The powers of the transformation of the complex plane into itself given by $(z, w) \mapsto (az, \bar{a}w)$, $a = e^{2\pi i/q}$, form a group – the binary $q$-gon group. Prove that all polynomials invariant with respect to this group can be expressed in terms of $X = z^q$, $Y = w^q$, $Z = zw$, and that the orbit variety coincides with the surface $XY = Z^q$ in complex three-space. Prove that this surface is diffeomorphic to the zero level set of the simple function $A_{q-1}$ of three complex variables.

**91.** Prove that the orbit variety of the action of the binary tetrahedral (octahedral, icosahedral) group on the complex plane coincides with the zero level surface of the function $E_6$ ($E_7$, $E_8$) of three complex variables.

**92.** A collection of smooth submanifolds passing through the origin is called simple, if all nearby collections are exhausted by a finite list (up to a diffeomorphism of a neighbourhood of the origin). Find all simple collections on the plane and in three-space.

**93.** The critical point 0 of a smooth function $f(x, y)$ is called a simple boundary singularity (on the plane with boundary $x = 0$), if all nearby functions are exhausted by a finite list (up to a diffeomorphism of a neighbourhood of the origin preserving the line $x = 0$). Prove that the simple critical points of a function of two complex variables are exhausted by the list

$$B_k = x^k + y^2 \quad (k \geq 2); \qquad C_k = xy + y^k \quad (k \geq 3),$$
$$F_4 = x^2 + y^3$$

(the equation of the boundary being $x = 0$).

128

# References

*To the Preface*

The papers of Thom, Mather, Morin and others have been gathered together in the collection of (Russian) translations: Singularities of differentiable mappings. Mir, Moscow 1968, 268 p.

Articles discussed in the preface:

G. N. Tyurina: The topological properties of isolated singularities of complex spaces of codimension one. Izv. Akad. Nauk SSSR, Ser. Mat. 32 (1968), 605–620 (English translation: Math. USSR, Izv. 2 (1968), 557–571).

J. F. Nye, J. H. Hannay: The orientations and distortions of caustics in geometrical optics. Optica Acta 31 (1984), 115–130.

Yu. V. Chekanov: Caustics in geometrical optics. Funkts. Anal. Prilozh. 20:3 (1986), 66–69 (English translation: Funct. Anal. Appl. 20 (1986), 223–226).

On Thom's conjecture:

R. Thom: Topological models in biology. Topology 8 (1969), 313–335.

J. Guckenheimer: Bifurcation and catastrophe, in: Proceedings of the International Symposium in Dynamical Systems, University of Bahia (Salvador, 1971). Academic Press, New York 1973, 95–109.

B. A. Khesin: Bifurcation of singular points of gradient dynamical systems. Funkts. Anal. Prilozh. 20:3 (1986), 94–95 (English translation: Funct. Anal. Appl. 20 (1986), 250–252).

B. A. Khesin: Bifurcations in gradient dynamic systems, in: Itogi Nauki Tekh., Ser. Sovrem. Probl. Mat., Novejshie Dostizh. (Contemporary Problems of Mathematics, Newest Achievements) 33. VINITI, Moscow 1988, 113–155 (English translation: J. Sov. Math. 52 (1990), 3279–3305).

Montel on singularities:

P. Montel: Sur les méthodes récentes pour l'étude des singularités des fonctions analytiques, in: Trudy I. Vsesoyuznogo S"ezda Matematikov (Kharkov, 1930). ONTI NKTP SSSR, Moscow–Leningrad 1936, 36–57 (also published in Bull. Sci. Math., II. Sér. 56 (1932), 219–232.).

## To Chapters 1–5

An extensive bibliography can be found in the following sources:

T. Poston, I. Stewart: Catastrophe Theory and its Applications. Pitman, London–San Francisco–Melbourne 1978, 491 p.

V. I. Arnol'd, S. M. Gusein-Zade, A. N. Varchenko: Singularities of Differentiable Maps, vol. I. Nauka, Moscow 1982, 304 p. (English translation: Birkhäuser, Boston–Basel–Stuttgart 1985, 382 p.).

V. I. Arnol'd, S. M. Gusein-Zade, A. N. Varchenko: Singularities of Differentiable Maps, vol. II. Nauka, Moscow 1984, 336 p. (English translation: Birkhäuser, Boston–Basel–Berlin 1988, 492 p.).

E. C. Zeeman, B. W. W.: 1981 Bibliography on catastrophe theory. University of Warwick, Coventry 1981, 73 p.

V. I. Arnol'd: Singularities of systems of rays. Usp. Mat. Nauk 38:2 (1983), 77–147 (English translation: Russ. Math. Surv. 38:2 (1983), 87–176).

Itogi Nauki Tekh., Ser. Sovrem. Probl. Mat. (Contemporary Problems of Mathematics) 22. VINITI, Moscow 1983, 244 p. (English translation: J. Sov. Math. 27:3 (1984)).

Itogi Nauki Tekh., Ser. Sovrem. Probl. Mat., Novejshie Dostizh. (Contemporary Problems of Mathematics, Newest Achievements) 33. VINITI, Moscow 1988, 236 p. (English translation: J. Sov. Math. 52:4 (1990)).

V. I. Arnol'd (ed.): Dynamical Systems V. Itogi Nauki Tekh., Ser. Sovrem. Probl. Mat., Fundam. Napravleniya (Contemporary Problems of Mathematics, Fundamental Directions) 5. VINITI, Moscow 1986, 284 p. (English translation: Encycl. Math. Sci. 5. Springer-Verlag, Berlin–Heidelberg–New York, to appear 1992).

V. I. Arnol'd (ed.): Dynamical Systems VI. Itogi Nauki Tekh., Ser. Sovrem. Probl. Mat., Fundam. Napravleniya (Contemporary Problems of Mathematics, Fundamental Directions) 6. VINITI, Moscow 1988, 256 p. (English translation: Encycl. Math. Sci. 6. Springer-Verlag, Berlin–Heidelberg–New York, to appear 1992).

V. I. Arnol'd (ed.): Dynamical Systems VIII. Itogi Nauki Tekh., Ser. Sovrem. Probl. Mat., Fundam. Napravleniya (Contemporary Problems of Mathematics, Fundamental Directions) 39. VINITI,

Moscow 1989, 256 p. (English translation: Encycl. Math. Sci. 39. Springer-Verlag, Berlin–Heidelberg–New York, to appear 1993).

V. I. Arnol'd (ed.): Theory of Singularities and its Applications. Adv. Sov. Math. 1. American Mathematical Society, Providence, R. I. 1990, 333 p.

V. I. Arnol'd: Bifurcations and singularities in mathematics and mechanics, in: Theoretical and Applied Mechanics. Proceedings of the Seventeenth International Congress held in Grenoble, August 21–27, 1988. North-Holland, Amsterdam–New York 1989, 1–25.

S. G. Gindikin (ed.): Operator Theory in Function Spaces. Review lectures of the XIIIth All-Union School held at Kujbyshev (USSR), October 6–13, 1988 (in Russian). Izdatel'stvo Saratovskogo Universiteta, Kujbyshevskij Filial, Kujbyshev 1989, 223 p.

J. M. T. Thompson: Instabilities and Catastrophes in Science and Engineering. John Wiley & Sons, Chichester 1982, 226 p.

The first paper on the theory of singularities:

H. Whitney: On singularities of mappings of Euclidean Spaces. I. Mappings of the plane into the plane. Ann. Math., II. Ser. 62 (1955), 374–410.

Textbooks:

Th. Bröcker, L. Lander: Differentiable Germs and Catastrophes. Lond. Math. Soc. Lect. Note Ser. 17, Cambridge University Press, Cambridge 1975.

M. Golubitsky, V. Guillemin: Stable Mappings and Their Singularities. Grad. Texts Math. 14, Springer-Verlag, New York–Heidelberg–Berlin 1973, 209 p. (but see the corrections in the Russian translation: Mir, Moscow 1977, 290 p.).

R. Gilmore: Catastrophe theory for scientists and engineers. John Wiley & Sons, New York–Chichester–Brisbane–Toronto 1981, 666 p.

J. Bruce, P. Giblin: Curves and Singularities. Cambridge University Press, Cambridge–London–New York 1984, 222 p.

Discussions about catastrophes:

R. Thom: Topological models in biology. Topology 8 (1969), 313–335.

R. Thom: Stabilité Structurelle et Morphogénèse. W. A. Benjamin, Reading, Mass. 1972.

R. Thom: Catastrophe Theory: Its present state and future perspectives, in: Dynamical Systems – Warwick 1974. Lect. Notes Math. 468, Springer-Verlag, Berlin–Heidelberg–New York 1975, 366–372.

E. C. Zeeman: Catastrophe theory: A reply to Thom, in: Dynamical Systems – Warwick 1974. Lect. Notes Math. 468, Springer-Verlag, Berlin–Heidelberg–New York 1975, 373–383.

E. C. Zeeman: Catastrophe Theory. Selected Papers, 1972–1977. Addison-Wesley, Reading, Mass. 1977.

J. Guckenheimer: The catastrophe controversy. Math. Intell. 1 (1978), 15–20.

H. J. Fussbudget, R. S. Znarler: Sagacity theory: a critique. Math. Intell. 2 (1979), 56–59.

### To Chapter 2

M. L. Gromov, Ya. M. Ehliashberg: Construction of a smooth mapping with prescribed Jacobian. I. Funkts. Anal. Prilozh. 7:1 (1973), 33–40 (English translation: Funct. Anal. Appl. 7 (1973), 27–33).

Ya. M. Ehliashberg: On singularities of folding type. Izv. Akad. Nauk. SSSR, Ser. Mat. 34 (1970), 1110–1126 (English translation: Math. USSR, Izv. 4 (1970), 1119–1134).

### To Chapter 6

Poincaré's thesis (1879):

H. Poincaré: Sur les propriétés des fonctions définies par les équations aux différences partielles, in: Œuvres de Henri Poincaré, Tome I. Gauthier-Villars, Paris 1951, XLIX–CXXIX.

The thesis contains, among other things, a versal deformation theorem for zero-dimensional complete intersections (Lemme IV on page LXI) and the method of normal forms.

Andronov's work on the theory of structural stability and on bifurcation theory was already presented in the paper:

A. A. Andronov: Mathematical problems of the theory of self-oscillations (in Russian), in: I Vsesoyuznaya konferentsiya po kolebaniyam. GTTI, Moscow–Leningrad 1933, 32–72 (reprinted in Andronov's Collected Works, Moscow 1956, pp. 85–124.).

His 1939 paper (jointly with E. A. Leontovich) contains an investigation of both types of bifurcation involving the birth of a cycle: the local type (the cycle is born out of an equilibrium position) and the nonlocal type (the birth of a cycle out of a separatrix loop). See:

A. A. Andronov, E. A. Leontovich: Some cases of the dependence of limit cycles on parameters (in Russian). Uch. Zap. Gor'kovskogo Gos. Univ. No. 6 (1939), 3–24.

A. A. Andronov, [A. A. Witt], S. Eh. Khajkin: Theory of Oscillations. Fizmatgiz, Moscow 1937 (English translation: Princeton University Press, Princeton 1949) (in later editions it is stated that the name of the second author was omitted "owing to a tragic mistake". In the 1949 English edition, the authors' names were rendered as A. A. Andronow and C. E. Chaikin. Authors' names are of course not left out of books by accident. As the date of publication and the omission of his name suggest, Witt was executed in the Stalin purges; that was the "tragic mistake".)

The work on exponential divergence of trajectories is summed up in:

D. V. Anosov, Ya. G. Sinaj: Some smooth ergodic systems. Usp. Mat. Nauk. 22:5 (1967), 107–172 (English translation: Russ. Math. Surv. 22:5 (1967), 103–167).

S. Smale: Differentiable dynamical systems. Bull. Am. Math. Soc. 73 (1967), 747–817.

E. N. Lorenz: Deterministic nonperiodic flow. J. Atmospheric Sci. 20 (1963), 130–141.

Applications of exponential divergence of trajectories to the theory of hydrodynamic instability are described in:

V. I. Arnol'd: Sur la géométrie différentielle des groupes de Lie de dimension infinie et ses applications à l'hydrodynamique des fluides parfaits. Ann. Inst. Fourier 16,1 (1966), 319–361.

V. I. Arnol'd, B. A. Khesin: Topological methods in hydrodynamics. Annu. Rev. Fluid Mech. 24 (1992).

The papers cited in the text on bounds for the dimension of attractors:

Yu. S. Il'yashenko: Weakly contracting systems and attractors of the Galerkin approximations of the Navier-Stokes equations (in Russian). Usp. Mat. Nauk 36:3 (1981), 243–244.

Yu. S. Il'yashenko, A. N. Chetaev: On the dimension of attractors for a class of dissipative systems. Prikl. Mat. Mekh. 46 (1982), 374–381 (English translation: J. Appl. Math. Mech. 46 (1982), 290–295).

Yu. S. Il'yashenko: Weakly contracting systems and attractors of the Galerkin approximations of the Navier-Stokes equations on a two-dimensional torus (in Russian). Uspekhi Mekh. (Adv. in Mech.) 5:1/2 (1982), 31–63.

A. V. Babin, M. I. Vishik: Attractors of partial differential evolution equations and estimates of their dimension. Usp. Mat. Nauk. 38:4 (1983), 133–187 (English translation: Russ. Math. Surv. 38:4 (1983), 151–213).

Bogdanov's theorem was first announced in the survey:

V. I. Arnol'd: Lectures on bifurcations and versal families. Usp. Mat. Nauk 27:5 (1972), 119–184 (English translation: Russ. Math. Surv. 27:5 (1972), 54–123).

The proofs were published in:

R. I. Bogdanov: Bifurcation of the limit cycle of a family of plane vector fields. Tr. Semin. Im. I. G. Petrovskogo 2 (1976), 23–35 (English translation: Sel. Math. Sov. 1 (1981), 373–387).

R. I. Bogdanov: Versal deformation of a singularity of a vector field on the plane in the case of zero eigenvalues. Tr. Semin. Im. I. G. Petrovskogo 2 (1976), 37–65 (English translation: Sel. Math. Sov. 1 (1981), 389–421).

The cases of symmetries of order 2, 3 or ≥ 5:

V. K. Mel'nikov: The qualitative description of resonance phenomena in nonlinear systems (in Russian). Prepr. OIYaF P. 1013, Dubna 1962, 1–17.

E. I. Khorozov: Versal deformations of equivariant vector fields for the cases of symmetry of order 2 and 3 (in Russian). Tr. Semin. Im. I. G. Petrovskogo 5 (1979), 163–192.

Symmetry of order 4:

V. I. Arnol'd: Loss of stability of self-oscillations close to resonance and versal deformations of equivariant vector fields. Funkts. Anal. Prilozh. 11:2 (1977), 1–10 (English translation: Funct. Anal. Appl. 11 (1977), 85–92).

A. I. Nejshtadt: Bifurcations of the phase pattern of an equation system arising in the problem of stability loss of selfoscillations close to 1:4 resonance. Prikl. Mat. Mekh. 42 (1978), 830–840 (English translation: J. Appl. Math. Mech. 42 (1978), 896–907).

F. S. Berezovskaya, A. I. Khibnik: On the bifurcation of separatrices in the problem of stability loss of auto-oscillations near 1:4 resonance. Prikl. Mat. Mekh. 44 (1980), 938–943 (English translation: J. Appl. Math. Mech. 44 (1980), 663–667).

Delayed loss of stability:

M. A. Shishkova: Examination of a system of differential equations with a small parameter in the highest derivatives. Dokl. Akad. Nauk SSSR 209 (1973), 576–579 (English translation: Sov. Math., Dokl. 14 (1973), 483–487).

A. I. Nejshtadt: Asymptotic investigation of the loss of stability of an equilibrium upon slow passage of a pair of eigenvalues through the imaginary axis (in Russian). Usp. Mat. Nauk. 40:5 (1985), 300–301.

A. I. Nejshtadt: On delay of loss of stability under dynamic bifurcations. I. Differ. Uravn. 23 (1987), 2060–2067 (English translation under the title: Persistence of stability loss for dynamical bifurcations. I. Differ. Equations 23 (1987), 1385–1391).

A. I. Nejshtadt: On delay of loss of stability under dynamic bifurcations. II. Differ. Uravn. 24 (1988), 226–233 (English translation under the title: Persistence of stability loss for dynamical bifurcations. II. Differ. Equations 24 (1988), 171–176).

Cascades of doublings:

A. P. Shapiro: Mathematical models of competition (in Russian), in: Management and Information 10. Far-Eastern Scientific Centre of the Academy of Sciences of the U.S.S.R., Vladivostok 1974, 5–75.

R. M. May: Biological populations obeying difference equations: stable points, stable cycles, and chaos. J. Theoret. Biol. 51 (1975), 511–524.

M. Feigenbaum: Quantitative universality for a class of nonlinear transformations. J. Stat. Phys. 19 (1978), 25–52.

P. Collet, J.-P. Eckmann: Iterated Maps on the Interval as Dynamical Systems. Prog. Phys. 1, Birkhäuser, Basel–Boston–Stuttgart 1980, 248 p.

Bifurcations of codimension two:

G. Zholondek (H. Żołądek): On the versality of a family of symmetric vector fields in the plane. Mat. Sb., Nov. Ser. 120(162) (1983), 473–499 (English translation: Math. USSR, Sb. 48 (1984), 463–492).

H. Żołądek: Bifurcations of certain family of planar vector fields tangent to axes. J. Differ. Equations 67 (1987), 1–55.

Some recent papers:

V. I. Arnol'd: Dynamics of intersections, in: Analysis, et cetera: Research Papers Published in Honor of Jürgen Moser's 60th Birthday. Academic Press, San Diego 1990, 77–84.

V. I. Arnol'd: Dynamics of complexity of intersections. Bol. Soc. Bras. Mat. 21:1 (1990), 1–10.

V. I. Arnol'd: Majoration of Milnor numbers of intersections in holomorphic dynamical systems. Preprint 652, University of Utrecht, Utrecht April 1991, 1–9 (to appear in: Modern Topological Methods in Mathematics, Papers Published in Honor of J. Milnor's 60th Birthday. Publish or Perish, Austin 1992.).

## To Chapter 7

The finiteness theorem is proved in:

L. V. Levantovskij: Singularities of the boundary of the stability domain. Funkts. Anal. Prilozh. 16:1 (1982), 44–48 (English translation: Funct. Anal. Appl. 16 (1982), 34–37).

The simplest singularities are described in:

V. I. Arnol'd: Lectures on bifurcations and versal families. Usp. Mat. Nauk 27:5 (1972), 119–184 (English translation: Russ. Math. Surv. 27:5 (1972), 54–123).

## To Chapter 8

A different approach to the theory of metamorphoses of wave fronts and caustics is set forth in the article:

G. Wassermann: Stability of unfoldings in space and time. Acta Math. 135 (1975), 57–128.

It is interesting to note that an unfortunate choice of the point of view and the formulation of the problem led the author of this article to complicated answers in the simplest cases and hid from him the simple general laws, described in the works cited below, which govern the more complicated cases. Pictures of the metamorphoses of wave fronts in three-space first appeared in:

V. I. Arnol'd: Critical points of smooth functions, in: Proceedings of the International Congress of Mathematicians (Vancouver, 1974) vol. 1. Canadian Mathematical Congress, 1975, 19–39.

The theory of metamorphoses of caustics and wave fronts is set forth in the articles:

V. I. Arnol'd: Wave front evolution and equivariant Morse lemma. Commun. Pure Appl. Math. 29 (1976), 557–582.

V. M. Zakalyukin: Reconstructions of wave fronts depending on one parameter. Funkts. Anal. Prilozh. 10:2 (1976), 69–70 (English translation: Funct. Anal. Appl. 10 (1976), 139–140).

V. M. Zakalyukin: Legendre mappings in Hamiltonian systems (in Russian), in: Some Problems of Mechanics. MAI, Moscow 1977, 11–16.

A. B. Givental': Geometry, stability and symmetry. Current Science 59:21–22 (1990), 1052–1064.

A detailed exposition is to be found in V. M. Zakalyukin's thesis (Moscow State Unversity, Moscow 1978, 145 p., in Russian); see also:

V. M. Zakalyukin: Reconstructions of fronts and caustics depending on a parameter and versality of mappings, in: Itogi Nauki Tekh., Ser. Sovrem. Probl. Mat. (Contemporary Problems of Mathematics) 22. VINITI, Moscow 1983, 56–93 (English translation: J. Sov. Math. 27 (1984), 2713–2735).

V. I. Arnol'd: Singularities of Caustics and Wave Fronts. Math. Appl., Sov. Ser. 62, Kluwer Academic Publishers, Dordrecht–Boston–London 1990, 259 p.

The pictures of the metamorphoses of caustics first appeared in the first Russian version of the present book:

V. I. Arnol'd: Catastrophe theory (in Russian). Priroda, Issue 10 (1979), 54–63.

In the French translation by J.-M. Kantor (Matematica, May 1980, 3–20), these pictures were replaced by a page of R. Thom's comments.

The theory of bicaustics is presented in:

V. I. Arnol'd: Bifurcations of singularities of potential flows in a collision-free medium and metamorphoses of caustics in three-space. Tr. Semin. Im. I. G. Petrovskogo 8 (1982), 21–57 (English translation under the title: Evolution of singularities of potential flows in collision-free media and the metamorphosis of caustics in three-dimensional space. J. Sov. Math. 32 (1986), 229–258).

The results on bifurcations were announced at the Petrovskij seminar in the autumn of 1980 (see Usp. Mat. Nauk 36:4 (1981), p. 233), and the pictures of bicaustics appeared for the first time in 1981 in the first Russian edition of the present book. Some of these surfaces were studied in the work of Shcherbak and of Gaffney and du Plessis in 1982 (in Shcherbak's theory in the guise of unions of tangents to space curves).

The classification of the singularities of caustics and wave fronts up to dimension 10 was carried out in the article:

V. M. Zakalyukin: Lagrangian and Legendrian singularities. Funkts. Anal. Prilozh. 10:1 (1976), 26–36 (English translation: Funct. Anal. Appl. 10 (1976), 23–31).

and was corrected in § 21 of the book:

V. I. Arnol'd, S. M. Gusein-Zade, A. N. Varchenko: Singularities of Differentiable Maps. Volume I: The Classification of Critical Points, Caustics and Wave Fronts. Nauka, Moscow 1982, 304 p. (English translation: Birkhäuser, Boston–Basel–Stuttgart 1985, 382 p.).

The work on ice motion:

J. F. Nye, A. S. Thorndike: Events in evolving three-dimensional vector fields. J. Phys. A 13 (1980), 1–14.

Two recent references:

V. I. Arnol'd: Spaces of functions with moderate singularities. Funkts. Anal. Prilozh. 23:3 (1989), 1–10 (English translation: Funct. Anal. Appl. 23 (1989), 169–177).

V. A. Vasil'ev: The topology of spaces of functions without complicated singularities. Funkts. Anal. Prilozh. 23:4 (1989), 24–36 (English translation under the title: Topology of spaces of functions without compound singularities. Funct. Anal. Appl. 23 (1989), 277–286).

## To Chapter 9

E. M. Lifshitz, I. M. Khalatnikov: Investigations in relativistic cosmology. Adv. in Phys. 12 (1963), 185–249.

Ya. B. Zeldovich: Gravitational instability: an approximate theory for large density perturbations. Astron. Astrophys. 5 (1970), 84–89.

V. I. Arnol'd, S. F. Shandarin, Ya. B. Zeldovich: The large scale structure of the universe I. General properties. One and two-dimensional models. Geophys. Astrophys. Fluid Dyn. 20 (1982), 111–130.

V. I. Arnol'd: Bifurcations of singularities of potential flows in a collision-free medium and metamorphoses of caustics in three-space. Tr. Semin. Im. I. G. Petrovskogo 8 (1982), 21–57 (English translation under the title: Evolution of singularities of potential flows in collision-free media and the metamorphosis of caustics in three-dimensional space. J. Sov. Math. 32 (1986), 229–258).

V. I. Arnol'd: Some algebro-geometrical aspects of the Newton attraction theory, in: Arithmetic and Geometry II: Geometry. Prog. Math. 36, Birkhäuser, Boston 1983, 1–3.

S. F. Shandarin: Percolation theory and cellular structure of the universe (in Russian). Prepr., Inst. Prikl. Mat. Im. M. V. Keldysha Akad. Nauk SSSR, Mosk. No. 137, Moscow 1982, 1–15.

## To Chapter 10

L. N. Bryzgalova: Singularities of the maximum of a parametrically dependent function. Funkts. Anal. Prilozh. 11:1 (1977), 59–60 (English translation: Funct. Anal. Appl. 11 (1977), 49–51).

L. N. Bryzgalova: Maximum functions of a family of functions depending on parameters. Funkts. Anal. Prilozh. 12:1 (1978), 66–67 (English translation: Funct. Anal. Appl. 12 (1978), 50–51).

V. A. Vasil'ev: Asymptotic of exponential integrals, Newton's diagram, and the classification of minimal points. Funkts. Anal. Prilozh. 11:3 (1977), 1–11 (English translation: Funct. Anal. Appl. 11 (1977), 163–172).

V. I. Matov: The topological classification of germs of the maximum and minimax functions of a family of functions in general position. Usp. Mat. Nauk 37:4 (1982), 129–130 (English translation: Russ. Math. Surv. 37:4 (1982), 127–128).

V. I. Matov: Ellipticity domains of generic families of homogeneous polynomials and extremum functions. Funkts. Anal. Prilozh. 19:2 (1985), 26–36 (English translation under the title: Elliptic domains of general families of homogeneous polynomials and extreme functions. Funct. Anal. Appl. 19 (1985), 102–111).

I. A. Bogaevskij: Metamorphoses of singularities of minimum functions and bifurcations of shock waves of the Burgers equation with vanishing viscosity. Algebra Anal. 1:4 (1989), 1–16 (English translation: Leningr. Math. J. 1:4 (1990), 807–823).

## To Chapter 11

Davydov's classification was constructed in his thesis:

A. A. Davydov: Singularities in two-dimensional controlled systems (in Russian). Moscow State University, Moscow 1982, 149 p.

The results were partially announced in:

A. A. Davydov: Singularities of the admissibility boundary in two-dimensional control systems. Usp. Mat. Nauk 37:3 (1982), 183–184 (English translation: Russ. Math. Surv. 37:3 (1982), 200–201).

A. A. Davydov: The boundary of attainability in two-dimensional controlled systems (in Russian). Usp. Mat. Nauk 37:4 (1982), 129.

The proofs were published in:

A. A. Davydov: The boundary of the attainable set of a multidimensional control system. Tr. Tbilis. Univ. 232–233, Ser. Mat., Mekh., Astron. 13–14 (1982), 78–96 (English translation: Sel. Math. Sov. 5 (1986), 347–356)
(on the Hölder and Lipschitz continuity of the boundary).

A. A. Davydov: Normal forms of differential equations, not resolved with respect to the derivative, in a neighbourhood of a singular point. Funkts. Anal. Prilozh. 19:2 (1985), 1–10 (English translation under the title: Normal form of a differential equation, not solvable for the derivative, in a neighbourhood of a singular point. Funct. Anal. Appl. 19 (1985), 81–89).

A. A. Davydov: The normal form of slow motions of an equation of relaxation type and fibrations of binomial surfaces. Mat. Sb., Nov. Ser. 132(174) (1987), 131–139 (English translation: Math. USSR, Sb. 60 (1988), 133–141).

A. A. Davydov: Singularities of fields of limiting directions of two-dimensional control systems. Mat. Sb., Nov. Ser. 136(178) (1988), 478–499 (English translation: Math. USSR, Sb. 64 (1989), 471–493).

On Davydov's theorems see:

V. I. Arnol'd: Ordinary Differential Equations (3rd edition) (in Russian). Nauka, Moscow 1984, 272 (English translation: Springer-Verlag, Berlin–Heidelberg–New York 1992, 12, 334 pp.).

V. I. Arnol'd: Contact structure, relaxation oscillations and singular points of implicit differential equations, in: Geometry and Theory of Singularities in Nonlinear Equations: A Collection of Scientific Works. Voronezh University Press, Voronezh 1987, 3–8 (English translation in: Global Analysis – Studies and Applications III. Lect. Notes Math. 1334, Springer-Verlag, Berlin–Heidelberg–New York 1988, 173–179).

Singularities of convex hulls, the case of a surface in three-space:

V. M. Zakalyukin: Singularities of convex hulls of smooth manifolds. Funkts. Anal. Prilozh. 11:3 (1977), 76–77 (English translation: Funct. Anal. Appl. 11 (1977), 225–227).

Curves in three-space:

V. D. Sedykh: Singularities of the convex hull of a curve in $\mathbb{R}^3$. Funkts. Anal. Prilozh. 11:1 (1977), 81–82 (English translation: Funct. Anal. Appl. 11 (1977), 72–73).

The general case:

V. D. Sedykh: Singularities of convex hulls. Sib. Mat. Zh. 24:3 (1983), 158–175 (English translation: Sib. Math. J. 24 (1983), 447–461).

V. D. Sedykh: Functional moduli of singularities of convex hulls of manifolds of codimensions 1 and 2. Mat. Sb., Nov. Ser. 119(161) (1982), 223–247 (English translation: Math. USSR, Sb. 47 (1984), 223–236).

Singularities of the shadow of a convex surface:

C. O. Kiselman: How smooth is the shadow of a smooth convex body? J. Lond. Math. Soc., II. Ser. 33 (1986), 101–109.

V. D. Sedykh: An infinitely smooth compact convex hypersurface with a shadow whose boundary is not twice-differentiable. Funkts. Anal. Prilozh. 23:3 (1989), 86–87 (English translation: Funct. Anal. Appl. 23 (1989), 246–248).

I. A. Bogaevskij: Degree of smoothness for visible contours of convex hypersurfaces, in: Theory of Singularities and its Applications. Adv. Sov. Math. 1. American Mathematical Society, Providence, R. I. 1990, 119–127.

## To Chapter 12

Y. L. Kergosien, R. Thom: Sur les points paraboliques des surfaces. C. R. Acad. Sci., Paris, Sér. A 290 (1980), 705–710.

[The mistakes were partially corrected in the paper:

Y. L. Kergosien: La famille des projections orthogonales d'une surface et ses singularités. C. R. Acad. Sci., Paris, Sér. I 292 (1981), 929–932.]

O. A. Platonova: Singularities of the mutual disposition of a surface and a line. Usp. Mat. Nauk 36:1 (1981), 221–222 (English translation: Russ. Math. Surv. 36:1 (1981), 248–249).

O. A. Platonova: Singularities of projections of smooth surfaces. Usp. Mat. Nauk 39:1 (1984), 149–150 (English translation: Russ. Math. Surv. 39:1 (1984), 177–178).

O. A. Platonova: Projections of smooth surfaces. Tr. Semin. Im. I. G. Petrovskogo 10 (1984), 135–149 (English translation: J. Sov. Math. 35 (1986), 2796–2808).

E. E. Landis: Tangential singularities. Funkts. Anal. Prilozh. 15:2 (1981), 36–49 (English translation: Funct. Anal. Appl. 15 (1981), 103–114).

A more detailed exposition can be found in the theses of Platonova (Moscow State University, Moscow 1981, 150 p.) and Landis (Moscow State University, Moscow 1983, 142 p.).

V. I. Arnol'd: Singularities of systems of rays. Usp. Mat. Nauk 38:2 (1983), 77–147 (English translation: Russ. Math. Surv. 38:2 (1983), 87–176).

O. P. Shcherbak: Projectively dual space curves and Legendre singularities. Tr. Tbilis. Univ. 232–233, Ser. Mat., Mekh., Astron. 13–14 (1982), 280–336 (English translation: Sel. Math. Sov. 5 (1986), 391–421).

The proofs of the theorems on projections are based on the paper:

V. I. Arnol'd: Indices of singular points of 1-forms on a manifold with boundary, convolution of invariants of reflection groups, and singular projections of smooth surfaces. Usp. Mat. Nauk 34:2 (1979), 3–38 (English translation: Russ. Math. Surv. 34:2 (1979), 1–42).

A different approach to projections is presented in the book:

T. Banchoff, T. Gaffney, C. McCrory: Cusps of Gauss Mappings. Res. Notes Math. 55, Pitman, Boston–London–Melbourne 1982.

A survey on the singularities of projections:

V. V. Goryunov: Singularities of projections of full intersections, in: Itogi Nauki Tekh., Ser. Sovrem. Probl. Mat. (Contemporary Problems of Mathematics) 22. VINITI, Moscow 1983, 167–206 (English translation: J. Sov. Math. 27 (1984), 2785–2811).

See also:

V. V. Goryunov: Geometry of bifurcation diagrams of simple projections onto the line. Funkts. Anal. Prilozh. 15:2 (1981), 1–8 (English translation: Funct. Anal. Appl. 15 (1981), 77–82).

V. V. Goryunov: Projection of 0-dimensional complete intersections onto a line and the $k(\pi, 1)$-conjecture. Usp. Mat. Nauk 37:3 (1982), 179–180 (English translation: Russ. Math. Surv. 37:3 (1982), 206–208).

V. V. Goryunov: Bifurcation diagrams of some simple and quasihomogeneous singularities. Funkts. Anal. Prilozh. 17:2 (1983), 23–37 (English translation: Funct. Anal. Appl. 17 (1983), 97–108).

V. V. Goryunov: Projection and vector fields, tangent to the discriminant of a complete intersection. Funkts. Anal. Prilozh. 22:2 (1988), 26–37 (English translation: Funct. Anal. Appl. 22 (1988), 104–113).

### To Chapter 13

V. I. Arnol'd: Critical points of functions on a manifold with boundary, the simple Lie groups $B_k$, $C_k$, and $F_4$ and singularities of evolutes. Usp. Mat. Nauk 33:5 (1978), 91–105 (English translation: Russ. Math. Surv. 33:5 (1978), 99–116).

O. A. Platonova: Singularities in the problem of the quickest by-passing of an obstacle. Funkts. Anal. Prilozh. 15:2 (1981), 86–87 (English translation: Funct. Anal. Appl. 15 (1981), 147–148).

O. A. Platonova: Singularities of a ray system near an obstacle (in Russian). Prepr., Dep. VINITI 11 Feb. 1981, No. 647–81, Moscow 1981, 150 p.

V. I. Arnol'd: Singularities in variational calculus, in: Itogi Nauki Tekh., Ser. Sovrem. Probl. Mat. (Contemporary Problems of Mathematics) 22. VINITI, Moscow 1983, 3–55 (English translation: J. Sov. Math. 27 (1984), 2679–2713).

### To Chapter 14

The theory of Lagrangian singularities was founded in 1966. See:

V. I. Arnol'd: Characteristic class entering in quantization conditions. Funkts. Anal. Prilozh. 1:1 (1967), 1–14 (English translation: Funct. Anal. Appl. 1 (1967), 1–13).

L. Hörmander: Fourier integral operators, I. Acta Math. 127 (1971), 79–183.

V. I. Arnol'd: Integrals of rapidly oscillating functions and singularities of projections of Lagrangian manifolds. Funkts. Anal. Prilozh. 6:3 (1972), 61–62 (English translation: Funct. Anal. Appl. 6 (1972), 222–224).

V. I. Arnol'd: Normal forms for functions near degenerate critical points, the Weyl groups of $A_k$, $D_k$, $E_k$ and Lagrangian singularities. Funkts. Anal. Prilozh. 6:4 (1972), 3–25 (English translation: Funct. Anal. Appl. 6 (1972), 254–272).

See also:

J. Guckenheimer: Catastrophes and partial differential equations. Ann. Inst. Fourier 23,2 (1973), 31–59.

The theory of Legendre singularities first appeared in the book:

V. I. Arnol'd: Mathematical Methods of Classical Mechanics. Nauka, Moscow 1974, 432 p. (English translation: Grad. Texts Math. 60, Springer-Verlag, New York–Heidelberg–Berlin 1978 (2nd edition of the English translation 1989), 462 p. (2nd edition 508 p.)).

and in the report:

V. I. Arnol'd: Critical points of smooth functions, in: Proceedings of the International Congress of Mathematicians (Vancouver, 1974) vol. 1. Canadian Mathematical Congress, 1975, 19–39.

See also:

M. J. Sewell: On Legendre transformations and elementary catastrophes. Math. Proc. Camb. Philos. Soc. 82 (1977), 147–163.

J.-G. Dubois, J.-P. Dufour: La théorie des catastrophes. V. Transformées de Legendre et thermodynamique. Ann. Inst. Henri Poincaré, Nouv. Sér., Sect. A 29 (1978), 1–50.

On the open swallowtail see:

V. I. Arnol'd: Lagrangian manifolds with singularities, asymptotic rays, and the open swallowtail. Funkts. Anal. Prilozh. 15:4 (1981), 1–14 (English translation: Funct. Anal. Appl. 15 (1981), 235–246).

V. I. Arnol'd: Singularities of Legendre varieties, of evolvents and of fronts at an obstacle. Ergodic Theory Dyn. Syst. 2 (1982), 301–309.

A. B. Givental': Lagrangian varieties with singularities and irreducible $sl_2$-modules. Usp. Mat. Nauk 38:6 (1983), 109–110 (English translation: Russ. Math. Surv. 38:6 (1983), 121–122).

A. B. Givental': Varieties of polynomials having a root of fixed comultiplicity and the generalized Newton equation. Funkts. Anal. Prilozh. 16:1 (1982), 13–18 (English translation under the title: Manifolds of polynomials having a root of fixed multiplicity, and the generalized Newton equation. Funct. Anal. Appl. 16 (1982), 10–14).

The theorems of Givental' on submanifolds of symplectic and contact spaces first appeared in the first Russian edition of this booklet in 1981. They generalise the Darboux-Weinstein theorem (the difference being that in Givental's theorems the structures are restricted just to the vectors tangent to the submanifold). The Darboux-Weinstein theorem is proved in the article:

A. Weinstein: Lagrangian submanifolds and hamiltonian systems. Ann. Math., II. Ser. 98 (1973), 377–410.

On submanifolds of symplectic and contact spaces see also:

V. I. Arnol'd: A. B. Givental': Symplectic geometry, in: Dynamical Systems IV. Itogi Nauki Tekh., Ser. Sovrem. Probl. Mat., Fundam. Napravleniya (Contemporary Problems of Mathematics, Fundamental Directions) 4. VINITI, Moscow 1985, 5–139 (English translation in: Encycl. Math. Sci. 4. Springer-Verlag, Berlin–Heidelberg–New York 1990, 1–136).

V. I. Arnol'd: Singularities in variational calculus, in: Itogi Nauki Tekh., Ser. Sovrem. Probl. Mat. (Contemporary Problems of Mathematics) 22. VINITI, Moscow 1983, 3–55 (English translation: J. Sov. Math. 27 (1984), 2679–2713).

R. B. Melrose: Equivalence of glancing hypersurfaces. Invent. Math. 37 (1976), 165–191.

R. B. Melrose: Equivalence of glancing hypersurfaces. II. Math. Ann. 255 (1981), 159–198.

J. Martinet: Sur les singularités des formes différentielles. Ann. Inst. Fourier 20,1 (1970), 95–178.

R. Roussarie: Modèles locaux de champs et de formes. Astérisque 30 (1975).

M. Golubitsky, D. Tischler: An example of moduli for singular symplectic forms. Invent. Math. 38 (1977), 219–225.

A. B. Givental': Singular Lagrangian manifolds and their Lagrangian maps, in: Itogi Nauki Tekh., Ser. Sovrem. Probl. Mat., Novejshie Dostizh. (Contemporary Problems of Mathematics, Newest Achievements) 33. VINITI, Moscow 1988, 55–112 (English translation: J. Sov. Math. 52 (1990), 3246–3278).

V. I. Arnol'd: Surfaces defined by hyperbolic equations. Mat. Zametki 44:1 (1988), 3–18 (English translation: Math. Notes 44 (1988), 489–497).

V. I. Arnol'd: On the interior scattering of waves, defined by hyperbolic variational principles. J. Geom. Phys. 5 (1988), 305–315.

A. B. Givental': Lagrangian imbeddings of surfaces and unfolded Whitney umbrella. Funkts. Anal. Prilozh. 20:3 (1986), 35–41 (English translation: Funct. Anal. Appl. 20 (1986), 197–203).

V. I. Arnol'd: First steps in symplectic topology. Usp. Mat. Nauk 41:6 (1986), 3–18 (English translation: Russ. Math. Surv. 41:6 (1986), 1–21).

V. I. Arnol'd: Contact Geometry and Wave Propagation. Monogr. Enseign. Math. 34, L'Enseignement Mathématique, Université de Genève, Genève 1989, 56 p.

V. I. Arnol'd: Singularities of caustics and wave fronts Kluwer, 1991, Mathematics and its applications, Soviet series, vol. 62.

V. I. Arnol'd: Contact geometry: the geometrical method of Gibbs's thermodynamics, in: Proceedings of the Gibbs Symposium, Yale University, May 15–17, 1989. American Mathematical Society, Providence, R. I. 1990, 163–179.

The saying about the lark's crest is quoted by Plutarch: "But since, as it would seem, not only all larks must grow a crest, as Simonides says, but also every democracy a false accuser (sycophant) …"*.

## To Chapter 15

A more detailed exposition can be found in the following books:

J. Milnor: Singular Points of Complex Hypersurfaces. Ann. Math. Stud. 61, Princeton University Press, Princeton 1968, 122 p.

V. I. Arnol'd, S. M. Gusein-Zade, A. N. Varchenko: Singularities of Differentiable Maps. Volume II: Monodromy and asymptotics of integrals. Nauka, Moscow 1984, 336 p. (English translation: Birkhäuser, Boston–Basel–Berlin 1988, 492 p.).

V. I. Arnol'd, V. V. Goryunov, O. V. Lyashko, V. A. Vasil'ev: Singularity theory, in: Dynamical Systems VI. Itogi Nauki Tekh., Ser. Sovrem. Probl. Mat., Fundam. Napravleniya (Contemporary Problems of Mathematics, Fundamental Directions) 6. VINITI, Moscow 1988, 1–256 (English translation in: Encycl. Math. Sci. 6. Springer-Verlag, Berlin–Heidelberg–New York, to appear 1992).

E. Brieskorn: Die Milnorgitter der exzeptionellen unimodularen Singularitäten. Bonn. Math. Schr. 150, Math. Inst. der Universität Bonn, Bonn 1983, 225 p.

E. Brieskorn, H. Knörrer: Ebene Algebraische Kurven. Birkhäuser, Boston 1981, 964 p.

Papers on the icosahedron:

O. V. Lyashko: Classification of critical points of functions on a manifold with singular boundary. Funkts. Anal. Prilozh. 17:3 (1983), 28–36 (English translation: Funct. Anal. Appl. 17 (1983), 187–193).

---

* According to D'Arcy Thompson, the crested bird referred to by Plutarch is probably not the skylark (*Alauda arvensis*), which migrates to Greece, but rather the lark *Alauda cristata,* the commoner species in southern Europe, which has a more conspicuous crest than the skylark and is indigenous to Greece all year round (Transl. note).

O. P. Shcherbak: Singularities of families of evolvents in the neighborhood of an inflection point of the curve, and the group $H_3$ generated by reflections. Funkts. Anal. Prilozh. 17:4 (1983), 70–72 (English translation: Funct. Anal. Appl. 17 (1983), 301–303).


*To Chapter 16*

The passage quoted from Thom:

R. Thom: Catastrophe Theory: Its present state and future perspectives, in: Dynamical Systems – Warwick 1974. Lect. Notes Math. 468, Springer-Verlag, Berlin–Heidelberg–New York 1975, 372.

Quivers:

P. Gabriel: Unzerlegbare Darstellungen I. Manuscr. Math. 6 (1972), 71–103.

I. N. Bernshtejn, I. M. Gel'fand, V. A. Ponomarev: Coxeter functors and Gabriel's theorem. Usp. Mat. Nauk 28:2 (1973), 19–33 (English translation: Russ. Math. Surv. 28:2 (1973), 17–32).

L. A. Nazarova, A. V. Rojter: Polyquivers and Dynkin schemes. Funkts. Anal. Prilozh. 7:3 (1973), 94–95 (English translation: Funct. Anal. Appl. 7 (1973), 252–253).

A. Dlab, K. M. Ringel: Representation of Graphs and Algebras. Carleton Math. Lect. Note 8, Carleton University, Ottawa 1974.

Regular polyhedra:

F. Klein: Vorlesungen über das Ikosaeder und die Auflösung der Gleichungen vom fünften Grade. B. G. Teubner, Leipzig 1884, 260 p.

J. McKay: Graphs, singularities and finite groups, in: The Santa Cruz Conference on Finite Groups (Univ. California, Santa Cruz, Calif., 1979). Proc. Symp. Pure Math. 37, American Mathematical Society, Providence, R. I. 1980, 183–186.

Boundary singularities:

V. I. Arnol'd: Wave front evolution and equivariant Morse lemma. Commun. Pure Appl. Math. 29 (1976), 557–582.

G. Wassermann: Classification of singularities with compact abelian symmetry. Regensb. Math. Schr. 1, Fachbereich Mathematik der Universität Regensburg, Regensburg 1977, 284 p.

V. I. Arnol'd: Critical points of functions on a manifold with boundary, the simple Lie groups $B_k$, $C_k$, and $F_4$ and singularities of evolutes. Usp. Mat. Nauk 33:5 (1978), 91–105 (English translation: Russ. Math. Surv. 33:5 (1978), 99–116).

M. Golubitsky, D. Schaeffer: A theory for imperfect bifurcation via singularity theory. Commun. Pure Appl. Math. 32 (1979), 21–98.

D. H. Pitt, T. Poston: Determinacy and unfoldings in the presence of a boundary, 1978. (A mythical preprint, cited in Chapter 16 of Poston and Stewart's book *Catastrophe Theory and its Applications* (Pitman, London–San Francisco–Melbourne 1978)).

P. Slodowy: Simple Singularities and Simple Algebraic Groups. Lect. Notes Math. 815, Springer-Verlag, Berlin–Heidelberg–New York 1980, 175 p.

D. Siersma: Singularities of functions on boundaries, corners, etc. Q. J. Math., Oxf. II. Ser. 32 (1981), 119–127.

V. I. Matov: Singularities of the maximum function on a manifold with boundary. Tr. Semin. Im. I. G. Petrovskogo 6 (1981), 195–222 (English translation: J. Sov. Math. 33 (1986), 1103–1127).

V. I. Matov: Unimodal and bimodal germs of functions on a manifold with boundary. Tr. Semin. Im. I. G. Petrovskogo 7 (1981), 174–189 (English translation: J. Sov. Math. 31 (1985), 3193–3205).

I. G. Shcherbak: Duality of boundary singularities. Usp. Mat. Nauk 39:2 (1984), 207–208 (English translation: Russ. Math. Surv. 39:2 (1984), 195–196).

I. G. Shcherbak: Focal set of a surface with boundary, and caustics of groups generated by reflections $B_k$, $C_k$, and $F_4$. Funkts. Anal. Prilozh. 18:1 (1984), 90–91 (English translation: Funct. Anal. Appl. 18 (1984), 84–85).

I. G. Shcherbak: Boundary singularities with a simple expansion (in Russian). Tr. Semin. Im. I. G. Petrovskogo 15 (1990) (English translation will appear in J. Sov. Math.).

Nguyễn hữu Đức, Nguyễn tiến Đại: Stabilité de l'interaction géométrique entre deux composantes holonomes simples. C. R. Acad. Sci., Paris, Sér. A 291 (1980), 113–116.

G. G. Il'yuta: Monodromy and vanishing cycles of boundary singularities. Funkts. Anal. Prilozh. 19:3 (1985), 11–21 (English translation: Funct. Anal. Appl. 19 (1985), 173–182).

The groups $H_3$ and $H_4$:

O. V. Lyashko: Classification of critical points of functions on a manifold with singular boundary. Funkts. Anal. Prilozh. 17:3 (1983), 28–36 (English translation: Funct. Anal. Appl. 17 (1983), 187–193).

O. P. Shcherbak: Singularities of families of evolvents in the neighborhood of an inflection point of the curve, and the group $H_3$ generated by reflections. Funkts. Anal. Prilozh. 17:4 (1983), 70–72 (English translation: Funct. Anal. Appl. 17 (1983), 301–303).

V. I. Arnol'd: Singularities in variational calculus (in Russian). Usp. Mat. Nauk 39:5 (1984), 256.

V. I. Arnol'd: Singularities of ray systems, in: Proceedings of the International Congress of Mathematicians, August 16–24, 1983, Warszawa vol. 1. North Holland, Amsterdam–New York–Oxford 1984, 27–49.

A. N. Varchenko, S. V. Chmutov: Finite irreducible groups, generated by reflections, are monodromy groups of suitable singularities. Funkts. Anal. Prilozh. 18:3 (1984), 1–13 (English translation: Funct. Anal. Appl. 18 (1984), 171–183).

A. B. Givental': Singular Lagrangian manifolds and their Lagrangian maps, in: Itogi Nauki Tekh., Ser. Sovrem. Probl. Mat., Novejshie Dostizh. (Contemporary Problems of Mathematics, Newest Achievements) 33. VINITI, Moscow 1988, 55–112 (English translation: J. Sov. Math. 52 (1990), 3246–3278).

O. P. Shcherbak: Wavefronts and reflection groups. Usp. Mat. Nauk 43:3 (1988), 125–160 (English translation: Russ. Math. Surv. 43:3 (1988), 149–194).

Some additional recent references:

V. I. Arnol'd: Theory of Singularities and its Applications. Lezioni Fermiane, Accademia Nazionale dei Lincei, Scuola Normale Superiore, Pisa 1989. Cambridge Uiniversity Press, Cambridge 1991, 120 p.

V. I. Arnol'd, V. A. Vasil'ev: Newton's *Principia* read 300 years later. Notices Am. Math. Soc. 36 (1989), 1148–1154.

V. I. Arnol'd, V. A. Vasil'ev: Addendum to "Newton's *Principia* read 300 years later". Notices Am. Math. Soc. 37 (1990), 144.

V. I. Arnol'd: Huygens and Barrow, Newton and Hooke: pioneers in mathematical analysis and catastrophe theory from evolvents to quasicrystals. Sovremennaya Matematika dlya Studentov (Contemporary Mathematics for Students), 1. Nauka, Moscow 1989, 96 p. (English translation: Birkhäuser, Basel–Boston–Berlin 1990, 118 p.).

V. I. Arnol'd: Bernoulli-Euler updown numbers associated with function singularities, their combinatorics and arithmetics. Duke Math. J. 63 (1991), 537–555.

V. I. Arnol'd: Springer numbers and morsification spaces. Preprint 658, University of Utrecht, Utrecht April 1991, 1–18.

## To the Appendix

A more detailed analysis of the applications of the ideas of catastrophe theory before its advent is to be found in the article:

V. I. Arnol'd: Catastrophe theory, in: Dynamical Systems V. Itogi Nauki Tekh., Ser. Sovrem. Probl. Mat., Fundam. Napravleniya (Contemporary Problems of Mathematics, Fundamental Directions) 5. VINITI, Moscow 1986, 219–277 (English translation in: Encycl. Math. Sci. 5. Springer-Verlag, Berlin–Heidelberg–New York, to appear 1992, 207–264), where an appropriate bibliography is also listed.

See also:

D. Bennequin: Caustique mystique. Séminaire N. Bourbaki No. 634 (1984). Astérisque 133–134 (1986), 19–56.

## To the Conclusion

T. L. Saaty: Mathematical models of arms control and disarmament. Operations Research Society of America, Publications in Operations Research No. 14. John Wiley & Sons, New York 1968, 190 p.

**V. I. Arnol'd,** Steklov Mathematical Institute, Moscow, Russia

# Ordinary Differential Equations

Translated from the Russian by R. Cooke

1992. IV, 334 pp. 272 figs. (Springer-Textbook)
Softcover  ISBN 3-540-54813-0

There are dozens of books on ODEs, but none with the elegant geometric insight of Arnol'd's book. Arnol'd puts a clear emphasis on the qualitative and geometric properties of ODEs and their solutions, rather than on the routine presentation of algorithms for solving special classes of equations.

Vector fields and one-parameter groups of transformations are introduced right at the start and Arnol'd uses this "language" throughout the book. This approach allows him to explain some of the real mathematics of ODEs in a very understandable way. The text is also rich with examples and connections with mechanics. Where possible, Arnol'd proceeds by physical reasoning, using it as a convenient shorthand for much longer formal mathematical reasoning.

Following Arnol'd's guiding geometric and qualitative principles, there are 272 figures in the book, but not a single complicated formula.

This book is an excellent text for a course whose goal is a mathematical treatment of different equations and the related physical systems.

B4.02.011